Self Science

A Guide to the Mind and
Your Brain's Potential

Self Science

A Guide to the Mind and Your Brain's Potential

Dr. Mandy Wintink, PhD

IGUANA

Publisher: Greg Ioannou
Editor: Kate Unrau
Front cover image: Chang Baek
Front cover design: Ellen Yu
Images (interior): Chang Baek
Book layout design: Holly Warren

Library and Archives Canada Cataloguing in Publication

Wintink, Amanda J. (Amanda Jane), 1975-, author
      Self science : a guide to the mind and your brain's
potential / Mandy Wintink, PhD.

Issued in print and electronic formats.
ISBN 978-1-77180-167-6 (paperback).--ISBN 978-1-77180-168-
3 (epub).--ISBN 978-1-77180-166-9 (kindle)

      1. Brain--Popular works. 2. Mind and body--Popular
works. I. Title.

QP376.W56 2016   612.8'2   C2016-903405-4
                                           C2016-903406-2

This is an original print edition of *Self Science*.

Contents

SECTION II — THOUGHTS, ACTIONS, AND EMOTIONS ... 93

To Ashar, my son, my teacher

Preface

I experienced a profound moment one day while reading the newly republished *The Organization of Behavior: A Theory of Neuropsychology*, a seminal book by the famous Canadian psychologist and neuroscientist, Donald Hebb. One of my professors at Dalhousie University, Richard Brown, had spent years studying the history of Hebb as a local icon — Hebb was Nova Scotian — and had convinced Psychology Press to reissue the book in 2002. As I read Brown's foreword in my fresh copy, the thought occurred to me that a major piece of neuroscience was missing — the personal perspective, the brain's ability to consider itself. It wasn't just the organization of other people's behaviour that was important in neuroscience, but rather, the organization of our own unique behaviour.

In fact, that idea spun me into a highly concentrated, euphoric, creative, and enlightened moment where everything in my life seemed to make perfect sense. Everything I had been studying both in the laboratory and through my own self-reflective and introspective journey seemed to align in that moment to give rise to what I termed "neuropsychoidiology" — the study of one's own brain, mind, behaviour, emotions, and thoughts. I scrambled to capture the essence of this epiphany. I wrote, I graphed, and I strived to capture this concept.

In that moment I also felt like I had complete understanding of who I was and how I had come to be — emotionally, psychologically, physically, spiritually, and universally. I felt peacefully complete in this new-found self-awareness. But more importantly, it ignited in me a mission: To share neuroscience and psychology with others as a path of self-development, self-discovery, self-knowledge, and self-awareness.

This book is the culmination of the 13 years since that original reflection, during which I have worked with this idea and delivered pieces of it through various courses, lectures, blogs, articles, and a coaching practice. But it is also a culmination of the 40 years of my life. To me, these ideas, this self-awareness journey through psychology and neuroscience, and various other mind activities feel as though they have always been part of me.

The result of this thinking is an approach to self-awareness and a simple invitation to explore your minds and brains together with me so we can all use our brains better and be our best selves. I invite you into this realm of self-awareness because I believe the study of the brain is not limited to neuroscientists in a laboratory. I see anyone who is curious about the mind and brain as a neuroscientist by virtue of his or her having a brain to study. Scientists typically engage in their investigations by looking at pictures of brains, recording neurochemicals, measuring electrical signals, using humans and laboratory animals, studying behaviour, conducting surgeries, and reviewing case studies of damage, to name but a few methods. Neuroscientists, as we normally think of them, get one glimpse into the brain, but the puzzle of how we, our minds and our brains, work is bigger than any one scientist can handle so neuroscience needs our brains too. In science, we rarely deal with an individual, preferring instead to work with groups and averages so we can be more certain of our results. Although current evidence-based practices provide highly reliable information, they do not always account for what is happening to each individual — to you and to me — in the same way that stats like "2.3 children" do not relate to any one family. Gold-standard science ignores the individual for the sake of generalization, as it must. But I happen to think the individual is equally as important. I think YOU and I are important in the equation. We have knowledge, insights, and reflections that are worthy observations.

Neuroscience is growing, and more is learned about the brain every day. Neuroscience is not just about diseases and disorders. A tremendous amount of neuroscience knowledge can benefit everyone; for example, the way the brain works efficiently, how it thinks in ways that both help and harm us, how it fools and tricks us, and how we can use its power to benefit us.

With many recent discoveries in neuroscience, the application of neuroscience to our lives is a relatively new luxury. Here, some of that neuroscience knowledge is offered in order to benefit the everyday person.

Using this book, we will push to understand the brains and minds within ourselves and others, through self-reflection paired with current neuroscience understanding. To assist, I offer my own reflections, insights, epiphanies, and even shameful accounts of ignorance. I offer my own experiences as a doorway into what might feel like a scary pursuit. In my experience, the more I share, the more others feel able to look inside.

Oftentimes when we venture into ourselves, we feel vulnerable. We may be afraid of what we might find. My hope is that if you feel that fear, you will proceed anyway because it really is a fun adventure. My aim in divulging my own experiences is not to self-indulge, but to inspire self-exploration in others and show you that I'm willing to go there too.

In this book, I will also provide ways to practice getting to know our brains better. It is a lifelong study and a deeply fascinating one at that! Recently, I started experiencing some depression and deep sadness for no apparent reason. While reflecting on this for some time, I noticed a great deal about how my mind and brain were working. For example, any time I experienced depression, I quickly tried to push the feeling back down under the surface. I realized that I was much more comfortable living in my state of anxiety and high energy than I was in facing depression, apathy, or stagnation. I had been aware of my tendency for anxiety for quite some time, but I was much less aware of any bouts of depression. In this recent reflection, I realized I had actually been avoiding looking at the depression. With this realization I committed myself to self-observation.

This act of observing is an integral part of the scientific method; therefore, I invite you into the world of science through the simple act of observation. One of the things I miss most about being in the laboratory daily is the pure joy of observation and experimentation, but I continue to engage in it as often as I can. That same curiosity of a scientist exists within a pursuit of self-discovery. I encourage each of us to be a scientist, to make comparisons with the general, to experiment with our brains, to test different variables, and to test how those variables affect the way we think, feel, and behave. Observe your brain. Observe my brain. Observe other brains. But be mindful that observation is not the same as overanalysis. Instead, through observation we can reflect as if we were aiming to see things as they truly exist — inside, outside, and all around us. We can also use others to mirror or reflect back who we are.

This book is laid out to inspire self-discovery. Throughout the book there are opportunities for you to stop, think, reflect, and even take notes. Some of these opportunities will be formal invitations (for example, self-reflection prompts, exercises, and chapter assignments), but any time you feel compelled to stop, think, reflect, and take notes, please do so. Section I is meant to give some basic psychology knowledge as we see it through self-knowledge and self-awareness and in neuroscience, and to provide a glimpse into how powerful the mind actually is. Section II, on thoughts, actions, and emotions, is meant to go a bit deeper into each of those areas with explicit examples taken from daily life. The chapter on stress provides a basic understanding of both natural and dysfunctional stress. The chapter on positive thinking and optimism teaches how we think and how we can actually exert a significant level of control over our thinking patterns. By choosing strategies that promote optimism and positive thinking, we stand to do better in life. The chapter on success and failure explains failure from a biological perspective, but I also suggest that failures are natural and necessary for success. This section includes a chapter on motivation, curiosity, and passion to help you to understand basic driving forces, including what motivates internally and externally. Section III is devoted to understanding how to harness the power of the mind by exploring the brain and the mind through complementary chapters on goal setting, mindfulness meditation and yoga, intuition, and creativity.

Together, these topics will provide an introduction to elements of self-discovery that I think are most interesting and most useful as a guide to the mind and your brain's potential. I cannot promise that this journey will be easy or without fear, but I can guarantee that it will be interesting. I know there is an incredible power inside each of us that just needs to be realized, actualized, and materialized. This is what drives me, and I hope it will drive you too. I invite you to explore, discover, and learn now.

Section I

The Basics

Chapter 1

The Power of Self-Knowledge and Self-Awareness

Prioritizing self-knowledge and self-awareness is not for everyone; although, I would argue it should be. Some people don't seem to care about who they truly are or about the stuff within themselves. Others are just plain afraid of what they might find. Some people simply don't value self-awareness as a worthwhile pursuit. But indeed, knowing ourselves is highly valuable. If we know who we are, we can become thriving human beings who can contribute our skills, strengths, and virtues — which we are aware of — to the world. Without self-knowledge, how can we possibly place ourselves in the right relationships, in the right environments, in the right lifestyles, and among the right friends, or choose the right career path? We can't, at least not easily. When we are not living optimally in the right environments for us, we are drained and we become a drain on the system. We get sick, we don't volunteer to help others, and we fail to thrive in the deepest sense. It can also be a huge source of existential anxiety and depression. Without self-knowledge, we may also flounder without direction, or we might mistakenly follow someone else's direction instead. Without self-knowledge, we can't aim for a different or better future — for ourselves or for others. Without self-knowledge, we can't know what would make us happy or miserable, when we are being abused or loved, when we deserve promotions, when we hurt someone else, or when we have the skills necessary to jump to the next level of success. Self-knowledge is a prerequisite for living a good, productive, and meaningful life, which ultimately benefits us and those around us.

Self-reflection is also a powerful part of self-knowledge or "metacognition," as it is described in psychology, and a part of being human. In the words of psychologist Stephen Fleming, "This knack for reflecting on our thoughts is often viewed as a hallmark of the human mind. It is also a vital survival skill. Metacognition is how we identify our limitations and compensate for them. A student who thinks she is unprepared for a chemistry exam, for example, can devote an extra evening to brushing up on atomic orbitals. When you set an alarm to remind yourself of something you suspect you will forget or make a to-do list to keep track of the day's activities, metacognition has stepped in to save you from your own deficiencies."[1]

Throughout this book there will be self-reflection prompts to help you to gain this skill. And yes, it *is* a skill, or as psychologist Howard Gardner refers to it, "intrapersonal intelligence," an intelligence about who we are as individuals, within us.[2]

---

1. Stephen M. Fleming, "The Power of Reflection," *Scientific American Mind* 25, no. 5 (Sep/Oct 2014): 30–37.

2. You can find links to Gardner's work at howardgardner.com/multiple -intelligences.

We don't have to go very far to experience opportunities for self-knowledge and self-awareness. Every experience offers up an opportunity for us to learn more about who we are, what we want to do, what we should do, what we can do, and what our place is in this world, but only if we take the time to reflect and in the process gain self-knowledge. A relationship that ends terribly or smoothly gives us knowledge about how, or how not, to act in the future. A job interview that we either nail or bomb informs us about what to do differently or what aspects of ourselves to highlight or downplay. Even doing an entire degree in one discipline only to realize we are actually interested in something entirely different is helpful, no matter when we realize it. Ultimately, self-knowledge and self-awareness come down to knowing more about who we are, where we are going, and where we came from, so that we can orient ourselves to a better, brighter, and more meaningful future.

First, what is "self"? Given that there is no single agreed-upon definition of the self, and that we can consider it from many different angles, I encourage each of you to develop your own definition. This is your first lesson. What does "self" mean to you? Jot down some ideas then take a stab at a definition.

Self-Reflection Exercise

What is your definition of "self"?

Notes:

Biology and the Self: Where Do I Come from?

A big part of knowing who we are starts with knowing where we come from. One way to do this is to consider our development. From a biological perspective, we undergo a tremendous amount of physical development both before and after birth. From the time of conception, we are rapidly growing and developing our physical body, including our mental and intellectual capacity. Humans develop from the fusion of a single sperm and egg into a two-celled organism then into a four-celled organism then into an eight-celled organism, a sixteen-celled organism and so on. The internal organs start to develop quickly and then the nervous system starts to develop. Specific cells even start to beat rhythmically right in the location where the heart will be. Within a very short amount of time, only weeks, we begin to look vaguely like babies and continue to grow into the little people we become when we are born. We develop into miraculous humans from the fusion of two cells. Those two cells, as they continue to divide, are the cells that will become all other cells. They are stem cells. They give rise to every other cell type in the body. That, to me, is an incredible example of the human potential.

All of this remained relatively intellectual to me until the fall of 2014 when I was finalizing this manuscript and got pregnant for the first time. At 14 weeks pregnant we got an ultrasound, and I was completely awestruck at how much like a little human our baby looked. When the technician commented on how active the baby was it made me realize that this little person already had lots of the necessary elements for its own self-development including a brain, heart, and many sprouting organs that would help it respond to its environment in utero and afterwards.

When I learned I was pregnant, I have to admit that the realization I was incubating a "self" beyond mine was very surreal. Never before had I shared the same physical space with another this intimately. My body had always been mine. My concept of "self" evolved significantly as a result of being pregnant, indeed.

Genes

How we get from an egg and a sperm fusing starts with genetics. Many people are not interested in genetics because they haven't been taught it in a way that feels particularly relevant. However, when considering who/what/when/where/why we are the way we are, it's hard not to turn to genetics. It's also hard not to consider the interaction between genetics and environment, a concept well known in psychology as the nature-nurture debate. At this point, it's rarely much of a debate. Nature (our genes, inheritance, biology) and nurture (our environment, upbringing, and social circumstances) both interact in many ways to give rise to all that we become.

When we are conceived, a sperm (from the biological father) fuses with an egg (from the biological mother) bringing together two complementary sets of chromosomes (46 in total, 23 from mother and 23 from father). After the fusion, one fused cell divides into two cells, then into four, then eight, then sixteen, and so on. We continue to divide and divide and divide and eventually our cells differentiate into all the cells of our body to form organs like the heart and the brain. That development starts from blueprint instructions coded within our DNA.

DNA (deoxyribonucleic acid) is a double-helix structure discovered in the 1950s (relatively recently!) by biologists Francis Crick and James Watson, who later won a Nobel Prize for their achievements. Their discovery of the DNA double helix allowed us to better understand how inheritance is passed on from generation to generation.

Our DNA contains a simple code or alphabet, made up of four letters, which are known as nucleic acids (A for adenine, G for guanine, C for cytosine, T for thymine). These letters make up the "base pairs," or the rungs, of the double helix ladder. In the 1980s a large multi-researcher project was proposed to begin to sequence the entire human genome (that is, the letter sequence), consisting of an estimated 30,000 genes. The goal of the Human Genome Project was to determine the sequence of base pairs that make up human DNA and identify and map all of the genes of the human genome from both a physical and functional standpoint, the idea being that we would learn the exact basis of all humans.

Practically speaking, by understanding the sequence, we could then make sense of all the proteins that the DNA codes for. And proteins are the basis for our existence and all of our biological functions, used for making cells and organs, for chemical reactions in our body, launching protective immune responses, and even for forming memories.

The genes contained within our DNA are of interest for many reasons including understanding basic human functioning, susceptibility to disease, and inheritance of traits. There is also a level of genetics that includes how the environment influences our genes. We used to see DNA as being stable and unchanging, but we now know that our DNA sequence can be modified, essentially by making some parts of the sequence more or less "readable." The genes themselves seem to be rather stable; the activation and suppression of at least some genes appear to be modifiable. The study of the modifications of the genes is known as "epigenetics." Whereas genetics has to do with the study of the structure of the DNA sequence that codes for our genes, epigenetics has to do with genetic changes that change how the DNA sequence is read, translated, and ultimately expressed.

Epigenetics is a sort of imprinting, whereby the environment leaves its footprints on the DNA. It serves as a great model for understanding how both genes and environment ultimately interact to give rise to who we are. This happens through a process called "DNA methylation" (an addition of a methyl group to the DNA sequence) or "histone deacetylation" (a transfer of the acetyl group to coenzyme A). The consequence of these two separate mechanisms is either a suppression of gene activity, or the reverse effect if demethylation and histone acetylation occurs. During methylation, the DNA stays tightly wound and is therefore less able to express itself, resulting in a suppression or reduced gene expression. During acetylation, the chromatin is transformed into a more relaxed structure that is associated with greater levels of gene transcription (or gene expression).

Epigenetic changes, like cancer, have been implicated in a variety of nature-nurture debates. For example, methylation in specific DNA regions silences genes that normally serve to suppress tumour growth (known as tumour suppressor genes). Yes, this means that we have genes that are designed for tumours. When they are shut down, they lead to tumour growth and can lead to cancer. Alternatively, demethylation can also activate other genes known as "oncogene" and their activation also leads to cancer. With respect to environment, there is much evidence to suggest that diet can alter methylation. There are recent estimates that diet can prevent up to 80% of colorectal cancers. Altered DNA methylation in many cancers, not just colorectal cancer, shows promise as a biomarker for cancer susceptibility. However, alterations in DNA methylation are also reversible, further showing how the environment affects our nature.

Looking at the nature-nurture debate through the lens of epigenetics is just one way in which to show that we are a product neither of our nature nor of our environment. We are influenced by both. Who, what, why, where, and how we are is a result of so many different factors converging to this one moment in time to give rise to ourselves. This relationship is, in fact, so complex that as scientists we have very little grasp of it. The Human Genome Project was anticipated to shed much more light on the particulars of the relationship than it has. Instead, the project has opened up a bigger window to the complexity of not only gene-gene interactions but also the complexity of gene-gene-gene-gene-environment-environment-and-so-on interactions!

Self-Development

Birth itself is a minor point along our developmental timeline; we continue to grow rapidly once outside the womb as we had while in the womb. Parents, and anyone else fortunate enough to watch this growth, see the amazing feats of development little humans undergo within the first few days after birth and throughout our first year of life. We develop language and the ability to ask questions and remember how to get from A to B. We grow taller and wider, and we develop teeth and stronger immune systems. Eventually, in late adulthood, we grow shorter and skinnier and weaker, almost as if we are regressing.

Despite a longstanding interest in the "self," how it actually develops remains somewhat elusive to many of us, including psychologists and non-psychologists alike and neuroscientists. We use the term "self" daily when referring to our thoughts, behaviours, emotions, interactions with others, and so on. Like many concepts in psychology (for example, love, awareness, consciousness, intelligence) "self" is not clearly defined, although it may be operationally defined for the sake of scientific investigation. The same concept of "self" has been studied and debated in philosophy circles since we have had evolved brains sophisticated enough to ask the question. On the other end of the spectrum, in the outer neuroscience circles, the concept is actively avoided.

The sense of self begins to develop very early on, as far as we can tell. Even within the first five months as infants we seem to be able to differentiate ourselves from others.[3] For example, when given an opportunity to look at themselves in mirrors or on video, infants spend more time looking at themselves than looking at other infants,

---

3. Philippe Rochat, "Five Levels of Self-Awareness as They Unfold Early in Life," *Consciousness and Cognition* 12 (2003): 717–731.

even when all those pictured are dressed in the same outfits. By 18 months of age, infants are able to distinguish themselves from other infants. Babies spend lots of time looking at faces and recognize if something is out of place. We test this in the lab (or you can try this one at home) by placing a sticker on a child's head. If we place children with stickers on their heads in front of a mirror, they will reach for their own heads rather than toward the mirror in a gesture of self-recognition. This mirror test is something that most other animals don't ever pass, with the exception of some primates like chimpanzees. Also by 18 months, children are able to refer to themselves as "me." This sense of self continues to evolve over time with several milestones. However, the point at which a sense of self emerges and is complete (if ever) is a matter of discussion and scientific investigation.

Although some psychologists, like Freud, ignored the development of the self in adulthood, others, like Erikson, developed theories that included changes across the lifespan. For example, Erikson proposed that the development of love and intimacy occur primarily in early adulthood; generativity and care primarily in middle adulthood and into old age; and integrity and wisdom primarily in old age.4 Another psychologist, Kohlberg, described another aspect of the self — moral development — with three stages that align with how we personally regulate our behaviour. In the early stage (pre-conventional) we develop a sense of obedience and later a sense of individuality related to personal gain. In the second stage (conventional) we develop

---

4. Eric H. Erikson, *Childhood and Society* (New York: W.W. Norton & Company, 1993), 247–269.

interpersonal skills and then a sense of social order. It is not until the later (post-conventional) that we begin to develop moral reasoning that goes beyond "law and order," essentially recognizing the nuances and grey areas in determining right and wrong. At that point we develop an understanding of universal principles of right and wrong, despite what the "rules" might say. This later stage requires higher development beyond basic learning. It requires comprehension and questioning. It also allows us to make decisions that a computer cannot make using a simple algorithm.[5]

Positive psychology, a division of psychology put forth by Martin Seligman in 1998, also values self-development. In his Positive Psychology Manifesto,[6] Seligman describes the need to study how humans and communities thrive. Positive psychology represents many elements of personal growth in adulthood. In fact, since 1998, there has been a resurgence of interest into personal growth that allows humans to thrive, including research into meaning and purpose, life satisfaction, wellness, happiness, and gratitude, all of which were largely ignored in a discipline that favoured the study of the abnormal. As a result, self-development is more recognized as an important contributor to health, wellness, success, and the ability to thrive. I imagine that this type of evolution will continue for years to come as we collectively seek to evolve our psychological and neurological capacities as human beings. In each of these models, what becomes clear is that there is much to

---

5. "Can Robots Be Programmed to Be Ethical? UK Researcher Puts Isaac Asimov's Laws to the Test," *As It Happens*, CBC Radio (CBC, September 18, 2014).

6. Martin E. P. Seligman, and Mihaly Csikszentmihalyi, "Positive Psychology: An Introduction," *American Psychologist* 55, no.1 (2000): 5–14.

consider around how we and the "self" continue to evolve and develop throughout a lifetime.

Neurobiology of Self

In the study of the neurobiology of self, we have come to realize that many parts of the brain may be participating in our sense of self. For example, parts of the brain, if destroyed, impinge on the ability to recognize ourselves. Damage to areas within the parietal cortex,[7] for example, can give rise to a condition known as *hemineglect,* where damage to the brain disrupts a person's ability to sense parts of the body as his or her own. A person with *hemineglect* literally loses the cells that are responsible for receiving information from that the part of the body. An example of the consequence of losing this sense of self is that a person might have the sudden impression that someone else is also in her bed, failing to realize that the limb she sees is actually her own!

Another area of the brain known as the insular cortex seems to be involved in aspects of self-awareness. The insular cortex monitors activities like blood pressure and heartbeat, important aspects of self. This is a great example of how our brain aims to be self-aware.

Eastern Philosophy and the Self

In Western psychology, we often seek to plump up the self and the ego through confidence-building tactics and therapy. We want to feel strong, able, and independent.

---

7. See the figure of the brain and the cortical regions in the next chapter for the parietal cortex.

Self-esteem is big here. But Eastern philosophies (for example, Buddhism, Hinduism) seek to dissolve the self, or to detach from it. We seek to know the true self rather than a built-up self. The true self is spiritual and recognizes that we are interdependent with all beings and things. Through practices like yoga and meditation, we aim to see things as they truly are. These are deep and intentional self-awareness practices aimed to unveil who we really are. Much of the Western psychology thinking around the self ignores some incredible wisdom offered up through Eastern philosophies, or psychologies, as I like to call them. Many Westerners aim to bridge this disconnect between Eastern and Western thinking. My favourites include Stephen Cope, Michael Stone, Eckhart Tolle, and, arguably, Abraham Maslow.

Stephen Cope, author of the *Wisdom of Yoga* and *The Quest for the True Self* refers to yoga as "a methodological approach to the restoration of the full potential of the human being."[8] As a Western psychotherapist himself, he says that "Americans suffer inordinately with what therapists call problems of the self — an inability to self soothe; an inability to sustain a satisfying and cohesive sense of self over time; an inability to warmly love the self; an inability to maintain an ongoing sense of belonging and a deep sense of meaning and purpose in life." He refers to this — although it's not a term he coined himself — as "self-estrangement." He says, "we live cut off from a sense of our true deep mutual belonging and interdependence, and we suffer from a painful sense of separation — a separation from the life of the body; a separation from the

---

8. Stephen Cope, *The Wisdom of Yoga: A Seeker's Guide to Extraordinary Living*, (New York: Bantam Books, 2006).

hidden depths of life, its mystery and interiority; a separation from the source of our own guidance, wisdom, and compassion; and a separation from the life-giving roots of human community. Alienation from self is the entire focus of yoga philosophy."

Despite this separation — this alienation — yoga offers a path (or a practice) that brings us back home to our true self, our core, our soul, and our spirit. Cope says that through yoga, we find that "there is a great deal more of us to experience. More consciousness, more energy, more awareness, more equanimity, more life in the body, more connection with the mysteries of the soul. And there is that wonderful, haunting voice of the true self that calls to us, that keeps us company as we stride deeper and deeper into the world, determined to save the only soul we really can save."

In yoga, we also question the self or the "I" with which we closely identify. "Who is the Real Me, then?" asks Stephen Cope. He continues,

> What yogis observed is that we're composed of many patterns, and many patterns that do not even fit so well together. And none of the pieces it does have seem to have exclusive title to 'I.' It is partly the driven attempt to make all the pieces fit nicely together that keeps us so often at right angles to life. The self, as we like to think of it, does not actually exist. The 'self' is more circumscribed than we want it to be. The clinging to our ideas of self — to the 'me' — causes suffering. It involves the clinging to a nonexistent entity.

Cope's ideas are an interpretation derived from studying the ancient yoga text, the *Yoga Sutras* written down by

Patanjali. Sutra means "thread" and implies a threaded concept throughout the 196 points described by Patanjali. In the *Yoga Sutras*, Patanjali says (according to Cope's translation), "A succession of consciousnesses, generating a vast array of distinctive perceptions, appears to consolidate into one individual consciousness." This refers to the development of the self as the integrator of our patterns. Essentially then, "this 'self' is masquerading as a single entity." Side note: yoga, as we have come to know it here in North America, is vastly different than what it was originally about, 6,000 years ago when it first developed. The physical practice we most widely identify with here in the West is only a small part of the greater yoga philosophy. In fact, most of Patanjali's yoga sutras refer to the deeper spiritual and mental practice of yoga.

Buddhism is also an ancient philosophy (or psychology). It emerged about 2,500 years ago and also speaks to this "detachment" from the ego. Very simply, Buddhism maintains that there are four noble truths: 1) there is suffering; 2) the causes of suffering are our attachment, cravings, and ignorance; 3) there is a way to end the suffering; and 4) here is the path to do so. Understanding what attachment, cravings, and ignorance are is a lifelong practice, and that's what the path offers.

The Path of Self-Awareness

The Buddha (the enlightened guy whose teachings founded Buddhism) said there were 8,000 paths toward enlightenment, to the freeing of the self and becoming one with all. When I heard reference to these 8,000 paths, I wondered if he had been referring to the number of people on earth, suggesting that every path was unique and personal. If so, today the Buddha would say there are over seven billion paths! I'd like to think this is what he meant because there are many ways for each of us to become more self-aware or even reach the ultimate in self-awareness. In my opinion, the path any of us chooses can't possibly be wrong because it is the path the self has chosen.

In our journey, we may begin to question our place in this world early on in our life and spend much of our life seeking the answers to our questions. This quest and any resulting answers are all part of our path. We may also arrive at a place where we feel confident that we implicitly understand the natural way of this universe, a sort of faith in how it all works. This may be through believing in the Father, Son, and Holy Ghost, or in Allah, or Krishna, or Brahma, Vishnu, Shiva, or any other representation (and personification) of the spirit.

Our beliefs may also lead us to practice a certain technique for self-awareness. We may adopt a practice of meditation or yoga. When we practise, we are undergoing self-development. When we practise being a Good Samaritan we are undergoing self-development. When we practise compassion, empathy, or loving kindness, we are undergoing self-development. The practice of these techniques is more than just a physical practice. The practice is largely to see ourselves as who we truly are, to reveal the parts of ourselves that make us feel connected to other humans. The practice is also about coming to terms with the truth that our physical body will leave this world, and then about adopting a philosophy around what happens to the essence of us after our physical body dies. These are all forms of self-development.

This high-level development ultimately involves the dissolution of the self after years spent developing the self. The Buddhist perspective is to first distinguish between the personal self and the universal self, the former being the self that we come to know as "I" and the latter representing our natural truth. Buddhists refer to these two entities in a few ways, including the universal self and the personal self, the big self and the small self, the no-self and the self, or the spiritual self and the psychological self. This is similar to the way we conceptualize the self in yoga. We may have heard our yoga teacher say, "leave your ego outside," and perhaps we have wondered what they meant. This refers to coming in without expectations, without seeing ourselves as distinctly different from each other, without attachment to what we can or cannot do on that particular day, without a striving mentality, and without our armour of inflated self-esteem. Instead, we are invited to come as we are. Pure selfless.

Consider the words of S.N. Goenka (the guru who started the hundreds of *Vipassana* meditation centres around the world based on teachings of the Buddha):

> *Vipassana* is a way of self-transformation through self-observation. It focuses on the deep interconnection between mind and body, which can be experienced directly by disciplined attention to the physical sensations that form the life of the body, and that continuously interconnect and condition the life of the mind. It is this observation-based, self-exploratory journey to the common root of mind and body that

dissolves mental impurity, resulting in a balanced mind full of love and compassion.[9]

On this path, whatever is unknown about yourself must become known to you … every day you will penetrate further to discover subtler realities about yourself, about your body and mind. [10]

You have to do the work; no-one else can do it for you. With all love and compassion an enlightened person shows the way to work, but he cannot carry anyone on his shoulders to the final goal. You must take steps yourself, fight your own battle, work out your own salvation. Of course, once you start working, you receive the support of all the Dhamma forces, but still you have to work yourself. You have to walk the entire path yourself.[11]

The way in which many Buddhists or yogis aim to dissolve the self is through self-knowledge or self-awareness and consciousness, which comprises much of the physical practice of sitting meditation and yoga, complemented of course by taking this practice to real life. The goal of meditation is to see the self as the cause of suffering and unhappiness, and to dissolve it and allow the universal self to reign. The Buddhist ideas of dissolving the self are analogous to what some Western psychological theorists also aim for in conventional psychotherapy. For example, in Gestalt therapy (developed by Fritz Perls) the goal is to remove the dishonesties created by society and by parental influences and to come back to the true self.

---

9. "Vipassana Meditation," *Polskie Stowarzyszenie Medytacji Vipassana,* accessed July 5, 2016, https://www.dhamma.org/en/about/vipassana.

10. S. N. Goenka, *The Discourse Summaries* (Washington Pariyatti, 2000): 13.

11. Ibid. 19.

These dishonesties arise, for example, when someone tells us we are a certain way and we adopt that as part of our identity. I personally maintain several identities, the clearest being "athletic," "attractive," "female," and to some extent "smart." Those are identities that peers, society, professors, coaches, and my parents have imposed upon me. Without them, I feel lost, as though I don't know who I am. But from an Eastern psychology and Gestalt psychology perspective, these identities are fallacies. They aren't really who I am.

Unlayering the self is an important part of finding our ways in this world. Too often we respond to the expectations that come with those labels and identities, the expectations of others. These expectations are heavy though, and they hold us back. Another label I maintain for myself is "independent," which makes it rather difficult for me to ask for or accept help when it is offered. To let go of that part of my identity means feeling a loss of security. Not being independent, a woman, an athlete, leaves me untethered to anything concrete. But that's exactly what freeing the self does.

Studying the Self: Practices of Self-Awareness and Self-Knowledge

That moment with Donald Hebb's book was, I think, a moment of enlightenment, albeit one that did not last. In that moment I was filled with complete clarity, understanding, and awe of who I was, where I came from, and my place in this world. It also spawned the term I now use to refer to the pursuit of self-knowledge and self-awareness: neuropsychoidiology. "Neuro" refers to the nervous system, and more practically for our purposes, the brain. "Psych" refers to the mind, as in psychology, but more progressively as the elements of the mind that include thinking (cognitions), emotions (affect), and actions (behaviours). "Idio" pertains to one's own, as in idiosyncrasy, and it further describes the awareness of our own mind and brain. Finally, "ology" refers to the study of, as in we are studying our own mind and brain, our self. Neuropsychoidiology, or NPI, is our own reflections about ourselves. It is a highly personal and missing element of neuroscience.

By learning about our own brain and psychology, we'll be able to learn something deeper about ourselves. As an example of where NPI can manifest, the other day my partner, Mike, texted me after listening to a podcast (*Invisibilia*) about visual empathy. The woman interviewed spoke about how she has a difficult time describing her own needs and desires but is able to experience them through others. She feels as though she is being hugged when she watches other people embrace, for example, on television. Mike compared this to the false "pregnancy symptoms" we joked he was experiencing during my pregnancy. After listening to this podcast, he concluded that his pregnancy symptoms were legitimate and stemmed from his exceptional visual empathy and overactive "mirror neurons." Mirror neurons consist of a type of brain cell that seem to reflect a sense of sameness. For example, when we perform a certain act (like moving our arm) certain motor neurons are activated. More interesting, when that same action occurs in someone else (for example, I see you moving your arm) my mirror neurons are activated and seem to signal or "mirror" what you are doing but in my own brain, hence the name. These neurons have been considered possible roots for empathy because they seem to represent the neural code for "I feel ya." Mike now assumes that his mirror neurons are hyperactive, a state that gave rise to his mirrored pregnancy symptoms. Indeed, perhaps he is onto something. I have long said that Mike takes on other people's passions quite readily. He struggles to know what he is passionate about, instead often following other people's passions. It's possible that he is, in fact, experiencing some kind of visual empathy, a mirror-neuron hyperactivity that conflated with his own reactions. This causes him to feel pregnant and to develop passion for other people's passions. In any case, my response to him was "good neuropsychoidiology!"

Another good neuropsychoidiologist I know is my friend and colleague, Allison. Allison took the very first life-coaching course I offered in 2010, in which we studied neuropsychoidiology. Not long after moving to Ontario, Allison was inspired by knowledge of how we map our environments, involving an activation of a part of the brain known as the hippocampus. One day she told me she was going to have a "hippocampal workout." I laughed. I knew exactly what she was doing. She was off to explore her environment. I can't tell you how happy that statement made me. More information about mapping is described in the chapters on curiosity and creativity, but for now, "hippocampal workouts" and "visual empathy mirror neurons" serve as examples of how we can take knowledge of the brain and apply it right back onto ourselves to develop self-awareness and self-knowledge.

Chapter Assignment:
Self-Reflection Questions and Self-Science

The purpose of this assignment is to get you thinking about who you are and the nature of you as a "self." Ask yourself these questions:

> ➢ Who am I?

> ➢ How well do I know myself?

> ➢ How can I get to know myself better?

> ➢ What factors have been major influencers on my development or on my sense of self?

> ➢ What aspects of that self were present when I was young that are or are not present now?

Next pick a few people close to you and ask them some questions about yourself to see how what they say about you matches or differs from what you believe about yourself. Here are some examples of questions:

> ➢ Who am I?

> ➤ What am I best known for, in your mind?

> ➤ What are my best qualities?

Positive psychology has given rise to the science of personal development. Visit the University of Pennsylvania Authentic Happiness questionnaire centre for one excellent resource: https://www.authentichappiness.sas.upenn.edu/testcenter.
Consider taking the VIA Survey of Character Strengths, the Brief Strengths Test, PERMA, and the Satisfaction with Life Scale.

Notes:

Chapter 2

Neuroplasticity and Changing the Brain

As a neuroscientist, I have studied the relationship between the brain and the mind for some time now, and I can tell you that there is no clear consensus on what exactly that relationship entails. In my opinion, the mind is a function and the brain is a biological entity that allows that function. The mind includes our thoughts, behaviours, and emotions, as well as our sense of identity and self, our knowledge, our intelligence, our language, our common sense, our relationships, our patterns, and that which we think of as "I." The mind is a very dominant part of our existence.

The relationship between the brain and the mind is so intertwined that if we begin to study the mind, we are inevitably studying the brain. And if we begin to study the brain, we are inevitably studying the mind. Similarly, when we change one, we change the other. The mind is one of the things the brain does. The brain's job is to organize, interpret, and transmit information from both the internal world and the external world to the rest of its body and within the brain itself. In so doing, the brain becomes an interface and transmitter of information and energy (electrical, chemical, etc.).

What Is Neuroscience?

Often the term "neuroscience" is equated with "brain science," but that's only partially true. In fact, *neuroscience* is the study of the entire nervous system. The brain sits at the top of the body, processes all the information that comes in through the sensory neurons, and directs the body to act accordingly by sending signals out through motor neurons. Information comes in through our senses including the eyes, ears, nose, mouth, and skin. Through these units we detect all kinds of information: visual, auditory (hearing), olfactory (smell), gustatory (taste), and somatosensory (touch, body sensations). All of these sensory receptors acquire information and send it up toward the brain to be further processed and relayed to the appropriate brain area. For example, visual information gets coded exactly as it comes in through our eyes. It first gets split up, or deconstructed: colour information is sent to some areas of the brain, line information to others, and even movement to other brain areas. The brain then reconstructs this information to form a perception of what exactly is being seen. All of this work happens below our level of awareness. There are even areas of the brain that, when damaged, knock out our ability to recognize human faces.

The *nervous system* essentially governs all of the other systems of the body, including reproduction, digestion, cardiovascular circulation, immune, musculoskeletal, breathing, the "fight-or-flight" fear response, and a whole host of mind experiences not limited to things like emotions, social relationships, rewards, goals-directed behaviour, creativity, intuition, sleep and wakefulness, sexual behaviour and orientation, gender, learning, memory... In fact, the list goes on as scientific evidence mounts, showing more relationships between the nervous systems and psychology.

Experience Your Own Nervous System

Exercise 1

Collect any of the following foods you might have available, something from each category:

Sweet	raisins (or dried apricots or bananas)
Sour	diluted lemon or lime juice
Bitter	citrus peel, dandelion greens, tonic water, unsweetened chocolate, olives
Salty	salted nuts, chips, soda crackers, pickles, cucumber
Umami	soy sauce

Try each food, one at a time. Put it in your mouth and let yourself experience it slowly. Swoosh the food around so it hits all the parts of your tongue. Notice what you experience. Compare this to what you see in the image below.

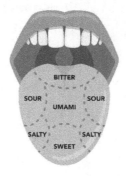

Figure 1. The tongue organized into five taste areas

Notes:

Exercise 2

Try to remain quiet and focus your attention on your breath for about 10 breaths.

Then let your attention drift to your heartbeat. *Can you feel it?*

Then let your attention drift to your belly. *Can you feel anything going on digestively?*

Then let your attention drift to your skin, any part that is exposed. *Can you feel the temperature?*

If you sit here long enough, notice how all of this is changing over time.

Notes:

Exercise 3

Sit quietly and think back to one of the most pleasant moments of your life. Perhaps it was when your child was born. Or when you got a puppy. When you got married or when you kissed the love of your life for the first time. Perhaps it was when you graduated or when got a promotion. Or maybe it was when you were at an amusement park with your best friend or when you learned to drive for the first time. Think of something that you can say was truly pleasurable, something that fills you with joy just thinking about the experience.

Let your mind wander with that memory for a few moments.

Notice, if you can, if you feel sensations anywhere within your body as you remember.

Notes:

In exercise 1 you were experiencing your taste buds, which when stimulated send signals to your brain so that you can actually experience those tastes as tastes. You might have also noticed that you experienced those tastes in distinct areas along the tongue. The tongue is mapped according to taste buds that either do or do not respond to each of our taste sensations: sweet, sour, bitter, salty, and umami (savoury). The information from our taste buds integrates in the brain with other information like past associations (maybe a taste reminds you of your grandmother's house and therefore integrates with your memory system), general positive experience (reaching the pleasure system of the brain), or disgust (reaching the vomit centres of the brain). In any case, by letting yourself experience each of those, you just experienced your own nervous system. If you happen to notice your tongue was different than what the map shows, you also just experienced neuropsychoidiology. You might also be experiencing some newer neuroscience in the field, which suggests the map may not be as originally described![12]

In exercise 2 you were experiencing parts of your *autonomic nervous system*, the division of the nervous system that controls heartbeat, breathing, digestion, and skin sensations, and regulates both our internal environment and information coming from our external environment. In exercise 3, you were experiencing emotions, the way they are both a psychological experience (one that you conjured up) and a physical experience. All of these exercises are portals to tapping into your own nervous system. Hopefully, what you learn next will give you some language to see things you weren't previously aware of. It's just like going to a wine tasting, beer tasting, or espresso tasting. Until you are

---

12. Greg Miller, "Sweet Here, Salty There: Evidence for a Taste Map in the Mammalian Brain," *Science* 333 (2011): 1213.

given the language to attach to your experience, you might not notice a difference in what you are tasting. Learning the language allows nuances to arise.

What you experience personally, internally, and through observation, in combination with what you learn about neuroscience in general will create a powerful foundation for you to understand the neural environment that gives rise to you, your thoughts, your emotions, your behaviours, your consciousness, your experience, your mind, and your wonderful uniqueness.

How the Brain Communicates and Governs: Structure and Function of the Nervous System

Let's review some of the pieces of the nervous system, starting with its anatomy (structure) and physiology (function). It's important to recognize that the brain is the governing agent for the entire body. Practically speaking, that means the brain is managing information from internal and external sources. It does this through the nervous system, which itself has several subdivisions. The brain "senses" what is going on inside and outside the body through *sensory* neurons and nerves. It commands the body to "move," or respond to sensation, through *motor* neurons. In other words, sensory neurons bring information in to be interpreted by the brain, and the brain makes those interpretations and responds by directing action or "output" through motor command neurons.

The nervous system collects, stores, filters, distributes, and makes sense of countless bits of information we respond to each day, each moment. The sensory and motor neurons converge in the part of *central nervous system* known as the spinal cord. It's easy to remember the central nervous system because the spine and the brain are in the centre of the body, hence the name. The brain is divided into a whole host of areas and networks of neurons forming specialized systems to manage certain behaviours. Cells are organized in groups based on similar function. These clusters of similarly functioning cells are referred to as *nuclei*. As an example, one system is called the *hypothalamic-pituitary-adrenal axis* (or HPA axis). The name refers to three clusters working together: the hypothalamus, the pituitary gland, and the adrenal glands. Together, this system helps to regulate the stress response, which will be discussed in the chapter on stress and self care. The other part of the central nervous system is the brain, where all the information converges and emerges. The other main division of the nervous system is the *peripheral nervous system*, which represents the connections with all experiences in the periphery, like sensing what is being touched by our baby toe and what is going on with our internal organs. The divisions of the peripheral nervous system are vast, but the most important thing to know is that information from the periphery eventually collects in the spinal cord and travels up to the brain (sensory pathways). Appropriate actions are then directed by the brain back to the periphery (motor pathways).

Within the peripheral nervous system, we have two other main divisions: the *somatic* and the *autonomic* nervous system. "Somatic" comes from the Greek word *soma*, meaning "body" and this system, aptly named, connects information about our body parts. The autonomic nervous system is further divided into the *sympathetic* and the *parasympathetic* nervous system, which will be covered further in the chapter on stress and self care. The sympathetic nervous system is the fight-or-flight arousal system and is responsible for all of those body sensations we feel when, for example, we are watching a scary movie. The opposite, the autonomic system, is the "rest and digest" system that works to calm us down after that scary movie. Both work together to prepare the body to deal with threats.

Neurotransmission — How Neurons Communicate with Each Other

In order for all of the systems to work together, they need some communicators. The neuron is a brain cell that serves as the basic unit of communication. Neurons communicate with each other in two main ways: chemically and electrically.

Chemical Communication of Information

We have all likely heard of the term "chemical imbalance", relating to mental illness. In many ways that descriptor has merit. Chemicals are part of how our neurons communicate with each other. Chemicals are even used to communicate with non-neurons, for example, to communicate with heart cells and muscles, and in digestion and reproduction through glands. There are several different classes of neurochemicals (or neurotransmitters). Some better-known examples include *serotonin* (pronounced sara-TONE-in), which is often thought to be imbalanced during depression and anxiety; *acetylcholine* (pronounced ah-see-tul-KO-lean), which is partially disrupted during Alzheimer's disease; and *dopamine* (pronounced DOPE-ah-mean), which is a reward chemical triggered by several types of addictive drugs like cocaine and amphetamines. GABA (*gamma-*Aminobutyric acid) is a neurochemical that, when increased, generally calms us down (for example, valium is a GABA drug). The opposite neurochemical to GABA is glutamate, which causes seizures when too much is present. Hormones can also function like neurochemicals. A good example of that is adrenaline, which most of us probably connect to the exhilarating feeling we get when we are running, jumping out of an airplane, riding a roller coaster, or watching a thriller movie. Adrenaline was discovered in the body and named as such, but the same chemical was discovered in the brain and called *epinephrine* (pronounced EPEE-nef-rin) before scientists realized they were studying the same chemical in different parts of the body. Many hormones are used to transmit messages in the brain, including sex hormones like estrogen and testosterone, and stress hormones like corticosteroids.

The messages that neurons need to communicate are relatively simple in principle. At the most basic level, they are binary: all or none, yes or no, stop or go. Each neuron has the potential to "fire" or not, upon stimulation. When a neuron fires, it releases neurochemicals. The chemicals released are the signals that travel between neurons and cause an adjacent neuron to fire or not by hitting specific receptors on that adjacent neuron. The right receptors have to be present for the message to be communicated. In fact, one way drugs interfere with our psychological experiences is by blocking the receptors, preventing messages from getting through. Sometimes that's good and sometimes that's not so good!

Let's consider a common drug as an example to describe this process better. Paxil, or fluoxetine, is a drug commonly used to treat depression and anxiety and is known as an SSRI, or selective serotonin reuptake inhibitor. The drug is selective for serotonin receptors and works by blocking the neuron's natural ability to recycle serotonin (known as "reuptake"), leaving the chemical (and its message) to linger longer. Little receptors also exist on the outside of the neuron communicating the original message. Those receptors help the neuron know if it's time to shut down the communication. By blocking (or inhibiting) those receptors, the original message stays active for a longer time. Selectively inhibiting the reuptake of a chemical like serotonin is one of the many ways drugs can interfere with our brain's messaging system.

This reuptake process keeps the system from diluting the message or sending too many of them. The process also keeps the environment "clean" so the system can respond to the next message quickly should another one come through. But under some conditions, like depression and anxiety, people don't seem to have enough of their natural chemical present.

Trying to rebalance the neurochemicals is a big part of many therapies in both mental health and neurodegenerative diseases (diseases where parts of the brain die and can no longer send or receive messages). For example, in Alzheimer's and Parkinson's disease, areas of the brain degenerate and can't release neurochemicals. With Alzheimer's disease there is a loss of the neurochemical acetylcholine, which normally sends messages to store memories. In Parkinson's disease there is damage to the cells that produce dopamine, resulting in a loss of dopamine, which normally sends messages related to motor movements. Some effects of these chemical imbalances are memory and spatial navigation deficits in Alzheimer's and loss of voluntary movements in Parkinson's.

Where Are Neurochemicals Located?

So far this description of chemical imbalance still leaves some unanswered questions. You may be wondering where exactly the chemicals are located. In fact, neurochemicals are found both in and between neurons. Neurons are separated from one another by small gaps called synapses. Chemicals are packaged in neurons and then moved from one end of the neuron to the other, where they can be released into the synapse. Once the neurochemicals are released, they float around and bump into their receptors on another neuron. If enough bump in, then a similar cascade of events is triggered in the next neuron, which typically results in another outpouring of neurochemicals from the end of that receiving neuron.

Chemicals function like a lock-and-key system with their receptors. When the right key finds the right lock, it opens a floodgate allowing the neuron on the other side of the synapse to fire. When that postsynaptic neuron (the neuron receiving the message on the other side of the synapse) fires, it too launches a cascade of events that typically triggers another chemical message to be conveyed. Neurons usually cluster with other similar neurons, so neurons within a cluster each respond to the same chemical. The many nuances to this communication allow a simple all-or-none communication process to increase in complexity and give rise to the range of complex experiences we and our brain have.

Where Do Neurochemicals Come From?

Most neurochemicals are built or derived from the foods that we eat. Food provides us with amino acids from the broken-down proteins. We use some of those amino acids to build new biological products. For example, serotonin is synthesized from an amino acid called tryptophan, which is found in a variety of foods including turkey and animal milk. Through a series of biochemical processes, tryptophan is eventually converted into serotonin (also known as 5-hydroxytryptamine, or 5-HT) and is then usable by neurons as the neurochemical serotonin. Because tryptophan is used to build serotonin, some medical practitioners (like naturopathic doctors) prescribe tryptophan (or another precursor, 5-hydroxytryptophan or 5-HTP) directly, as a treatment for depression. In theory, the net effect is similar to SSRIs: both increase serotonin availability in the synapse. There are pros and cons to that treatment, which I will not get into, but you could consult your naturopathic doctor or the scientific research for more information. Some medical practitioners might also suggest eating more foods that are high in the precursor of the neurochemical we want to build up, but there are some caveats to that method as well.

The conversion process for neurochemicals occurs in the neurons themselves, generally speaking. The details of this conversion are far beyond the scope of this chapter.

The important thing to understand is that generally neurochemicals are contained within the neuron in packages called synaptic *vesicles*. These vesicles eventually fuse with the lining at the end of the neuron and are released into the synapse.

Electrical Communication of Information

The neurochemicals remain packaged in vesicles until they receive the go-ahead to be released into the synapse. How the chemicals come to be released into the synapse is a matter of electricity, the second means of communication within the nervous system. When the chemical message is received, an electrical current is triggered within a neuron, which eventually results in more chemical being released at the other end of the neuron. The head of the neuron is the receiving end of the neuron, also known as the cell body or "soma." It receives messages when the chemicals find the right receptor. The chemicals bind to receptors on tree-like structures at the head of the neuron, called dendrites. When enough of the receptors along many dendrites are triggered, an electrical change in voltage amplitude, called an action potential, happens within the neuron. When we say a neuron fires, it means that the neuron has an action potential. The action potential is what moves the message internally down the neuron shaft (the "axon") and eventually leads to the release of the neurochemicals into the synapse, allowing the message to be transmitted from neuron to neuron. So between neurons, messages are sent using chemicals, but within a neuron, the message is transmitted with electrical currents and ions.

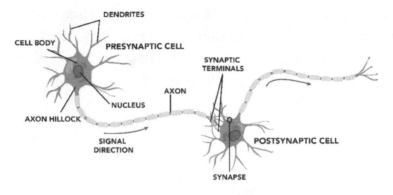

Figure 2. Two neurons

Measuring the electrical currents in the brain is one-way brain-recording devices are used to understand brain activity. One such device, an electroencephalograph (EEG), records and graphs electrical activity detected in the brain by using electrodes placed on the scalp. "Encephal" comes from the Greek word *kephalé*, meaning "head." That same root comes up in "encephalitis," which refers to an infection in the brain.

There are many, many important concepts from neuroscience that are worthy of reference, discussion, and mention. In fact, I own a book called *The Principles of Neural Science* that is the size of the Toronto Yellow Pages — an analogy that will soon be obsolete, but essentially it's massive! I also have another book that size on the hippocampus, one very specific cluster of brain cells that take up about one-sixteenth of the entire brain! The hippocampus cluster is known for its role in memory and, therefore, has been studied extensively. I have another book of similar size on the neurobiology of mental illness. Each of these books (and the many more that exist) offers a lot of information on how the brain works, yet not one author actually knows for sure. People say that we only use 10 per cent of the brain, but I think a better way of describing it is that we really only understand a fraction of how the brain works. Nonetheless, we can apply what we do know (or think we know) in wonderful ways as we continue to strive for more knowledge. And like I've said before, having a brain makes each of us a valid contributor to neuroscience. What's important for us is how the brain works (at least according to current knowledge), and particularly how to use the information that is most directly applicable to our lives. One of those is neuroplasticity and how the brain changes. As one 60-year-old student of mine said of the concept, "I now have hope that I can change." Indeed, she *can* change and so can her brain. And so can you and I. Hopefully this book will leave you feeling like you not only know your brain but are also empowered to cultivate the brain you want!

Neuroplasticity

We know a great deal more about the nervous system today than we did even 20 years ago, largely because neuroscience is young. At that time, most neuroscientists (and anyone who had studied biology of any sort) did not believe that new brain cells (*neurons*) could develop in the adult brain because of a long-held dogma that whatever you were born with was what you had for your lifetime. For a long time it was believed that our brains were relatively set in stone and lacked the ability to change. However, now we know that the brain has incredible potential to change and we have termed this *neuroplasticity*.

Technically, plasticity is the quality of being easily shaped or moulded, but that is exactly what the brain does. Plasticity is the reason why we develop from having a little neonatal brain to being able to read this text. Plasticity is the reason some of us learned to ride a bike, drive a car, or even do things we take for granted like walking. Plasticity is how we adjust and adapt to our environment and how we learn and remember. Plasticity is a major principle of neuroscience. Neuroplasticity refers to a plastic or changing nervous system and means that our nervous system changes, adapts, grows, dwindles, and evolves from conception through death. The brain is always changing and always engaged in neuroplasticity.

There are several ways in which the nervous system changes. Change occurs in the receivers of chemical messengers (for example, the dendritic spines and the receptors) both in number, size, and shape. Receptors themselves can either upregulate (sprout) or downregulate (wither away). These effects change the ways neurons connect and communicate. The speed at which neurons transmit their information down their shaft is also subject to change. When we improve the lining, a process called myelination, the transmission is faster, making communication better. Another kind of change is neurogenesis, a change in cell number. Neurogenesis literally involves making more brain cells and was only discovered about 20 years ago, which is recent as far as science is concerned.

Adult Neurogenesis — A New Discovery

Scientists have long known that neurogenesis occurred in the neonatal or newborn human brain. But we had assumed, until recently, that adult brains were incapable of generating new brain cells. We believed that we were born with a set number of brain cells, and that when those cells died, we had nothing with which to replace them. We use to believe that shortly after birth, at least for humans, the brain stopped producing new neurons. The changes that occurred after that point were believed to be related to changes in neuron connection and communication. In fact, many more neurons exist in our brains when we are born than are believed to be necessary. But after birth, a lot of pruning goes on, by virtue of a "use it or lose it" phenomenon. More neurons, and connections among those neurons, exist initially but many of those connections are lost when they are not supported or used. It's like having more than one road or path between points. With infrequent use, a city council might decide to no longer service the road less travelled. As a result, those roads get less and less use. Once out of commission, those paths are more difficult to resurrect. Difficult but not impossible.

The nature of pruning and plasticity helps explain why we experience a diminished ability to pick up new languages as we get older and why we might end up with accents. Different phonemes (the sounds within a given language) make up all languages, but not all phonemes are present in all languages. So as we continually hear one language, we lose sensitivity in the ability to hear the phonemes of other languages. Those other phonemes become "the road less travelled." As we age, we become less attuned to the particular phonemes of other languages and literally become deaf to them, at least in terms of what the brain processes. We don't hear words the same way that a native speaker hears them, and as a result, we unknowingly spit them back in conversation with those phonemes missing. Something similar happens with vision. If we are deprived of visual stimuli from birth, during a critical period when the visual region of the brain develops, we will lose the ability to have normal vision.

In the adult brain, pruning is less of an issue following the peak during infancy. It was long believed that once our neurons died, there were no new neurons to replace them. Then, about 20 years ago,[13] evidence to the contrary was discovered, and the concept (and very hot neuroscience field) of *adult neurogenesis*, that is, the generation of new neurons in the adult brain, was born. At present, this concept of adult neurogenesis is well accepted among neuroscientists and it is even making its way into the popular press. When I was an undergrad (1993–1998), we were still being taught that adult neurogenesis did not occur. By the time I was doing my PhD (2000–2005), however, I was studying adult neurogenesis in a very particular region of the brain, one that was known to generate new neurons. According to available research today, adult neurogenesis is localized to only a few discrete regions within the brain. During my PhD research, I was studying how the generation of new neurons

---

13. In actuality there was evidence back in the '60s but no one believed it. It was not until some work was published in 1994 that it began to hold water. For a review of all of this work, refer to Gerd Kempermann, *Adult Neuroscience 2* (New York: Oxford University Press, 2001).

contributed to the development of anxiety in laboratory rats. In some circumstances of abnormal brain activity (for example, in epilepsy) the brain started producing too many new neurons. These new neurons seemed to be causing havoc in the system, likely perpetuating the occurrence of seizures and allowing for the development of intense fear behaviour. This intense fear behaviour is somewhat analogous to the anxiety that can accompany certain forms of epilepsy. In that situation adult neurogenesis was not seen as a good form of neuroplasticity but, rather, one that went awry. When cells integrate poorly and form erratic connections within the system, it can lead to terrible communication among neurons.

During my post-doctoral fellowship research, I was also studying neurogenesis, but the purpose was to determine if we could harness the brain's natural capacity to regrow neurons in a way that could be helpful for neurodegenerative disorders (such as Parkinson's or Alzheimer's disease), where people suffer devastating losses of cells in discrete areas of the brain, resulting in significant behavioural, cognitive, and emotional changes.

Another popular area of study within the field of adult neurogenesis has to do with memory. One hotbed region where neurogenesis is known to occur — the hippocampus — is also known for its role in memory. Through several studies scientists have discovered that if neurogenesis is blocked with chemicals, memory is impaired. The opposite is also true. When the number of cells being born is increased, memory is enhanced. Some of this story has also unfolded through the study of people undergoing chemotherapy. Chemotherapy is a fairly crude way to prevent cell growth to halt the progression of cancer, but it also halts the growth of other desirable cells, including adult neurogenesis. Therefore, it wasn't surprising to learn that people undergoing chemotherapy suffer from memory problems and reduced adult neurogenesis, further supporting the hypothesis that the generation of new brain cells is required for memory formation. This is another example of how a changing brain contributes to the very personal experiences we have, including memory.

The idea that new brain cells could grow in the adult brain is a really recent concept. In the spring of 2013, I spoke to faculty members at a local university about the application of neuroscience in the classroom, and I mentioned the idea of adult neurogenesis. Following the lecture, a professor who teaches a developmental psychology class came up to me to ask if I was serious about the existence of adult neurogenesis. I had to tell her several times that I was indeed serious and had studied the phenomenon myself. She was astonished and embarrassed to say she was still teaching that adult neurogenesis was not possible. I assured her that she was not the only one who still held this dogma. A few years back, I was giving another lecture about how stress damages areas of the brain known to generate new brain cells when a medical doctor in the audience piped up to "correct me," indicating that adult neurogenesis did not exist. I assured her as well that this was a very real phenomenon being studied in many laboratories throughout the world. In fact, this area of study has exploded. When I first started my PhD in 2000, there were a mere handful of studies published on the phenomenon, such that I had read everything there was to know on the subject. But now, there are tens of thousands of papers published on adult neurogenesis. Needless to say, it was one of the hottest new discoveries in neuroscience in a long time.[14]

Neural Stem Cells

Stem cells have been quite popular in the media in the past decade or so, but many people still don't understand what they are. Stem cells are cells that have the capacity to become any kind of cell, a characteristic that makes them attractive and useful for regenerative purposes in the treatment of diseases and other damage. Imagine being able to create any body part you wanted in a human — organs, skin, or even limbs. The need for donor organs would change drastically. Stem cells have the potential to become needed body parts. Fetal stem cells have that potential because, by the nature of how we develop, they are already at that stage all that we will become. As already mentioned, at conception we begin as two cells (an egg and a sperm) that fuse to become four cells, which divide to become eight, then sixteen, then thirty-two, and eventually these cells start to differentiate into the various cells of the body. But until differentiation, they maintain the potential to become any part of the body. This potential is of significant interest when trying to repair humans, albeit controversial because we can currently only get them

---

14. Here's a list that some of my neuroscience colleagues and I recently vetted while arguing over a few missing pieces: dsc.discovery.com/tv-shows/curiosity/topics /10-amazing-advancements-in-neuroscience.htm.

from aborted fetuses. Stem cells do exist postnatally (after birth) too, for example, in the umbilical cord, and the preservation of the cord is becoming a popular new practice. Blood stem cells have also been found in bone marrow, making bone marrow transplants a treatment option for some diseases.

Neural stem cells also exist, in that there are certain brain cells that remain undifferentiated and have the potential to become any other brain cell, be they neurons of any particular type (for example, dopamine-producing or serotonin-producing) or even non-neuron brain cells, known as glial cells. Generally speaking, neural stem cells exist within areas where neurogenesis has been discovered; although, some scientists prefer to not make neurogenesis synonymous with neural stem cells.

In summary, neuroplasticity is a phenomenon that represents how and what we think, feel, do, say, and experience in our brain. Thinking, feeling, doing, saying, and all of our experiences are coded in our brain through changes that can be capitalized on for later use, as a form of learning. All of our experiences leave footprints on our brain (just like they leave footprints on our DNA) and those footprints take the form of distinct changes we lump under the term "neuroplasticity," or how the brain changes. Without the brain's ability to change, we would remain stagnant.

The Brain Is Organized

When the brain develops and brain cells grow, they organize themselves into function-specific regions, pathways, and structures. Major divisions of the brain include the cortex, subcortex, and brain stem structures.

Cortex

The cortex is a sheath of cells of similar function that envelop the structures beneath it, structures known as "subcortical" structures. Pictures of the brain, like the image included here, often show the cortex. The cortex is divided into four main areas, also called "lobes": frontal lobes, parietal lobes, temporal lobes, and occipital lobes.

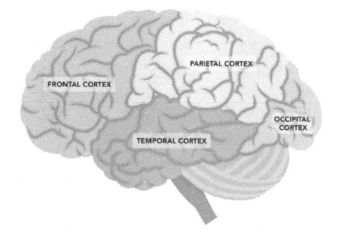

Figure 3. The four primary cortical regions of the brain: Frontal, Parietal, Occipital, and Temporal.

The occipital lobes involve the processing of vision, including the basics of colours, shapes, lines, depth, and movement, but also the perception of how those emerge into an image we understand. The temporal lobes are generally involved in processing functions like language, memory, and to some degree what and where we are. The parietal lobes are involved in processing awareness of space and our body.

The frontal lobes are the most fun to think about because they process several "executive functions" that humans use, often to our own destruction. These characteristics include things like planning ahead, logic, reasoning, rationality, working memory (holding something in mind), the ability to inhibit ourselves (from doing or saying stupid things), and impulsivity. The frontal lobes are almost like our CEO. It's easy to remember what the frontal lobes do once you learn that the frontal lobes are late to develop. They don't fully develop until we are in our late 20s or even early 30s according to some researchers. This should help us understand the behaviour of teenagers, who are often impulsive and don't think ahead about the consequences of their actions. They come by it honestly. The late development of the frontal lobes has become the basis for the "developing brain" theory that is gaining popularity in the legal system. When the brain is working really hard to keep things in mind, the frontal lobes are activated. When we are under the influence of alcohol, the frontal lobes are much less active. Sometimes other areas of our brain can also override our frontal lobes. For example, the amygdala (a brain structure that processes emotions) can sometimes be so active that it interferes with the rationality of the frontal lobes. To reinforce what the frontal lobes do, we can recall examples of these situations from our own lives. A well-functioning and well-trained brain can keep us in line. That training is the essence of the exercises throughout this book. Let's keep that in mind ... and keep our frontal lobes working.

The fact is, these executive functions seem to take up a lot of resources and are sometimes called "controlled processes." Controlled processes involve us actively and consciously trying to do something with the mind or brain. The opposite of controlled processes are "automatic processes," which happen behind the scenes, without much awareness. Examples of a controlled process is trying to remember all of the terms we have learned so far or trying to focus in a class that interests us very little. An example of an automatic process is seeing colour, because it happens without our awareness. Another example is any skill that we have learned that we no longer need to think about. The other day I was using a screwdriver to put together a table and chair set. I realized I normally use my right hand and never have to think about which way to move my hand. It is automatic. However, when I changed to my left hand, I did have to think about it. That was a controlled process. Over time, controlled processes do become automatic, and that's the way our brain likes it because automatic processing frees up valuable resources and allows us to learn more. More examples of processes that go from controlled to automatic include riding a bike, driving home along our routine route, or texting. Eventually, an efficient brain will adopt strategies to move those controlled processes to automatic ones, or, in other words, below our level of awareness. In this way, the brain is very smart.

Another important thing the frontal lobes are involved in is planning for the future and, along with portions of the temporal lobes, remembering the past. The strength of this feature is something that ancient psychologies of Buddhism and yoga address in their practices of meditation and *asanas*. This ability of ours to not be present in this moment by virtue of being lost in the future or the past is what those philosophies believe causes us suffering. But the aim of being present is working against a brain that has evolved to make good use of projecting ourselves into the past or the future. And, like all good developments, a word of caution is always warranted: have a strong bridge to go back and forth between. We will discuss this further in the mindfulness meditation and yoga chapter.

Subcortical Brain Regions

Below the sheath of the cortex lie many more regions of the brain, typically found in discrete clusters (brain structures) and pathways that channel information between structures up to the cortex and to brain structures in the back of the brain (for example, the brain stem). There are many brain structures involved in the processing of everything you can imagine about yourself including memory, hormonal regulation, body temperature, sexual behaviour, trust, motor control and balance, lying, addictions, and even how religious you might be. Structures work together by having similarly functioning cells all clustered together. For example, the hippocampus is a structure that has many like-minded cells working together to play a part in memory formation.

Subcortical structures also work with other like-minded structures. They organize themselves into systems. One example of a system is the "limbic system," known for its role in regulation and processing emotions. Seeing a bear, for example, activates the neurons in the retina of the eyes, and the neurons send information to the occipital lobes at the back of the brain. To get there, the information passes through a relay station in the middle of the brain, a structure known as the "thalamus." The thalamus directs different pieces of the information, for example, shape, size, contrast, and line orientation, to other structures specialized to receive that information. The amygdala is another brain structure that receives information from sight. The amygdala is primed to respond quickly to stimuli that it labels as "fearful." Adrenaline is released, sending a cascade of signals to engage the fight-or-flight system and prepare the body to move. The hippocampus, a structure already mentioned to be involved in memory and spatial awareness, takes the information it receives, processes it as a memory, and then provides input about whether the coast is clear or an escape route should be sought, while making a mental note of what to remember for the future should we see a bear again. The frontal lobes also help us decide if we need to escape and panic or inhibit our first response and calm down. Other structures at the back of the brain kick in to direct escape behaviour, should it be necessary. In a healthy brain system, all of these structures work together in concert with the primary goal of keeping you alive and out of danger.

Brain Stem Structures

Often we hear about the "reptilian brain." This part of the brain lies underneath the back of the head right where the neck meets the skull. The structures of the brain stem are responsible for vital functions such as breathing, heart rate, temperature, and vomiting. These reptilian structures are necessary to stay alive and that's about it. Their function isn't as sexy as that of the frontal lobes, the amygdala, or hippocampus, but they are important for survival. They are also often referred to as our "animal instincts," but that's not entirely true; many subcortical structures are involved in mating and other reproductive behaviours, as well as fear responses that are not found within the brain stem structures. That being said, the brain stem does contribute information to those survival behaviours too.

There are several major systems of the brain, all of which recruit different structures in order for the brain to do its job of governing the body and mind by processing information and directing behaviour, thoughts, and emotions. Rather than describe each of those, we will explore them if and when they become relevant during the chapters that follow. But generally speaking, all systems extend beyond the organization described. For example, being able to respond to an oncoming car requires communication among brain stem structures that alter heart rates (to push more oxygen to our muscles so we can move quickly) and structures that help us decide quickly (and automatically) if we are in danger. We also need structures to calm us down after so we don't remain in a hyper-alert and high-panic state. We need to calm down so we can become aroused again should the need arise. Although it sometimes comes across in the media as if certain features of our existence (like happiness or love) might take place only within a certain area of the brain, it's better to think of the brain as an entire system. The brain is one system within the bigger nervous system. The nervous system is made up of various cells that work together in clusters or structures, and those structures work together in their respective systems. Those systems are partially regulated and interpreted by activity at the cortical level and partially regulated by downstream systems that support basic functions like breathing and heartbeat. Together, the entire brain with all of its 100 billion neurons and other cells, are there to make us think, act, emote, and generally survive and reproduce. This system is part of how the simple all-or-none communication can give rise to the complexity of who we are, how we think, what we know, and how we behave. The ability of the brain to change, modify, adapt, and learn through neuroplasticity is yet another nuance the brain makes use of to create such complicated existences like our own.

Blaming Our Brains

Knowing all of these details about how the brain works may be interesting in and of itself, but let's consider this in context. Imagine you committed a crime but don't totally remember it. You were, perhaps, in a fit of rage because you came home to find your spouse in bed with another person. In hindsight you know that it was wrong, but at the time you were blinded by rage (hyperactivity in the amygdala and parts of the brain stem) and barely felt conscious while doing it (reduced activity in the frontal and parietal cortex). Perhaps the head injury that you suffered from the car accident a few months before contributed to your rage. Perhaps it wasn't the car accident but, rather, your upbringing, the horrific experiences you endured because of an abusive parent. Perhaps you inherited experiences from your mother or father (discussed in the chapter on stress). Could you blame your brain?

In fact, neuroscience is being used more and more in the legal system, particularly during pretrial determination of competency, during trials, and during sentencing. Neuroscience is making its way into the legal system so furiously that if a defendant does not undergo neuropsychological testing, a defence lawyer could be charged with ineffective council, a serious claim that is traditionally difficult to prove.

The use of neuroscience in the justice system is increasing in a variety of ways that were described in a very enlightening, philosophical, and intriguing talk by Nita Farahany at the Society for Neuroscience annual meeting in San Diego in 2014. Here are a few examples of how it has been used:

The Developing Brain Theory

This claim is being used as a blanket "truth" for teenagers and young people because the neuroscience is so compelling with respect to behaviours like impulsivity and decision-making. I am reminded of an accident in Ontario where a 19-year-old died car-surfing and a 16-year-old driver was charged with criminal negligence causing bodily harm and dangerous driving causing bodily harm. What struck me about that incident was that I remember returning home one "fun" evening with a bloody and swollen head and leg because of a similar incident where I, too, had been flung from the back of a car. I didn't know any better. And I assume neither did the Ontario teens, all because our brains simply were acting out as any developing frontal lobe brain might.

Reduced Monoamine Oxidase

Monoamine oxidase (MAO) is an enzyme that breaks down serotonin and naturally clears the synapses, keeping serotonin in balance and the neurochemical communication in check. MAOa is a gene that regulates the MAO enzyme, and its expression is blunted in some males. With that blunted gene expression, normal communication using serotonin is altered. In combination with severe childhood maltreatment like abuse, this results in males who are more prone to aggression and more likely to end up in the criminal justice system. The question, legally and scientifically, is whether these individuals are responsible for crimes committed as a result of negative early childhood experiences that were catalyzed by genes they inherited from their parents. How much are they to blame? And how much should be blamed on their genetic predispositions and horrible upbringings?

Drug and Alcohol Abuse

Drug and alcohol abuse is now often considered a medical disease. This has been helpful for people seeking treatment and in reducing the stigma associated with addictions. Drug and alcohol abuse has also been medicalized in the justice system as a rationale for why people might get high and commit crimes associated with such abuse. Crime is a by-product of their disease. Some defense lawyers have also argued an "unconscious brain" predisposed their clients toward drug- and alcohol-related crimes.

Brain Injuries

Brain injuries from such things as car accidents, physical violence, or other physical trauma, can render a person "never the same." Phineas Gage is the classic case; he was a man who was shot right through his brain with an iron rod during railway construction. He was never the same. Before his injury he was respected as a great foreman. But after his injury, he could no longer manage his crew, could not relate to humans the same way, and could not regulate his emotions. How would he fare in today's society? Imagine if he committed a crime when under the influence of his rage. What if he came home and found his spouse in bed with someone else and was unable to control himself because his frontal lobes were disconnected from his amygdala. If so, can we blame his brain?

There are many ways we may be affecting our brains without considering the consequences on who we are. Increased incidents of concussions in sports, for example, should be of serious concern. Several reports of depression and even suicide have made their way into the media with concussions as the likely cause. Are people with injured brains to blame if they do lash out and harm themselves?

Stress also damages the brain, including its memory systems and decision-making processes. The evidence is far from scarce. Are those with stressed brains to blame when they lash out or have a moment of blind rage? Am I to blame when I snap at someone because of a crazy day? Or when I'm pregnant? What about when we are on the receiving end? Should we blame? Or do we accept apologies? If so, does that mean we accept that circumstances (or the brain) are to blame?

In December 2012 Adam Lanza shot and killed twenty children and six adults in Sandy Hook Elementary School. Information about his medical and school history reported signs of severe anxiety but no overwhelming evidence for what drove him to massacre that day. The report also stated "significant mental health issues." Perhaps a brain scan would have been useful, but even without that, do we think his brain was to blame? Perhaps his MAOa gene was blunted and perhaps he suffered some unknown physical abuse as a child. If so, would that grant him more compassion?

The effect of this emerging neuroscience as it relates to legalities is forcing us to think about accountability. One of the benefits of my studying neuroscience is that I have a tolerance and compassion for human behaviour that stems from a biological rationale. In many ways, this begs the question of free will or lack thereof. What if we are a product of our brains, how they develop, and the experiences that shape them? I believe we are a product of our experiences and those experiences leave footprints in our DNA and in our brain. These footprints alter who we are from that moment on, in both huge ways and subtle ways, but nonetheless we are changing with each moment of our existence, and we are changing the existence of our offspring. But even if you don't believe that entirely, what if there is even an ounce of truth to it? If so, as Nita Farahany suggests, we as a society must consider the ramifications, most importantly with regards to our justice system. Our system relies on punishment and paying the price for our crimes. But is punishment the best way to deal with our brains? Would efforts directed toward rehabilitation be better served? Our brains are plastic; we know that. How plastic, we don't know. But if we shape our brains, should we not work towards reshaping them in a way that promotes societal integration? What happens to the brains of people who spend years in prison? Does that do anything to alleviate the problem that anxiety, isolation, abuse, or shame has created in the first place? It is something to think about, and I encourage you to do so.

Self-Directed Neuroplasticity

We can blame our brains if we want to for all the bad things that happen. But we can also flip that concept around and take a positive approach. If we now believe that our brains are responsible for who we are, what we think, how we act, and what we do AND if we believe that our brains are changeable, then we have to accept that we can, at least to some degree, direct our own fate and that of our brain. The concept of self-directed neuroplasticity was introduced by Dr. Jeffrey Schwartz. In 2003 he published *The Mind and the Brain: Neuroplasticity and the Power of Mental Force*,[15] in which he describes how we play an active role in influencing our brain's function by deciding where to focus our energy and attention. That is what we shall do throughout this book.

---

15. Jeffrey M. Schwartz and Sharon Begley. *The Mind and the Brain: Neuroplasticity and the Power of Mental Force* (New York: Harper Perennial, 2002).

Chapter Assignment:
Self Science — Getting to Know Your Brain

The purpose of this assignment is to help you discover more about how your brain works. To do this, we can employ the scientific process to conduct our own study of our brain. Although this experiment would not meet all of the criteria that rigorous scientific publications adhere to, it can provide you with some information regarding your own life.

Come up with a hypothesis that you want to test about your brain. A hypothesis is a testable educated guess. How do you think your brain works? What are you curious about in terms of your own brain?

Plan some methods for studying this hypothesis. If you wanted to test to see if that hypothesis is correct, how would you go about testing it? Could you introduce a change in your life in some way and then look to see if that change makes a difference? For example, if you believe that your brain works better when you get eight hours of sleep, how would you test that? One option would be to simply track each night's hours slept, record the quality of that sleep, and compare that to observations of how well your brain works each day. Perhaps you would observe your memory, your social interactions, your crankiness. You can come up with whatever measures you want.

Determine what kind of data you will collect. Related to the methods, you will want to very clearly identify what you are trying to measure; for example, the number of sleep hours and memory. Choose something of interest to you and something that allows you to more clearly state whether or not your hypothesis is correct.

Collect the data. This is the fun part. Do your experiment, record results, and keep track of what you are trying to observe and measure!

Analyze and interpret the data. You can compare your results over time or as one variable changes (sleep hours) and its effect on the other variable (memory). For now, unless you are a scientist with statistical background, just eyeball your data and see if you can determine a pattern in your brain and behaviour.

Draw a conclusion. Consider what it means in the big picture and whether you want to make any changes as a result of your discovery. Remember, it's your brain and your data, so make the conclusions relevant to you and not general to others.

<u>Notes:</u>

Chapter 3

The Power of the Mind

Several years ago I was tagged in a Facebook photo by someone I hadn't seen in years, so my initial reaction was somewhat apprehensive as I wondered what the photo would be. When I clicked on the notification, I was directed to a photo of three people. At first, I couldn't find myself, oddly enough. Then I realized that the one standing right in middle, a thin, muscular girl with a sports bra on, was me. Although I saw myself, I didn't recognize it as me. I was flooded with emotions. I felt sadness, defeat, and disgust because my body no longer looked like that, or so I thought. I stared at this photo for probably about 20 minutes confused. Why did I no longer look like this? What was wrong with me?

Then I began to remember my state of mind at the time when I did look like that. I remembered several important things. First, I was in lots of physical pain at the time. I was playing a lot of Ultimate Frisbee without warming up (because I was working too much and "didn't have time" for that). I would walk home from the field in pain, hunched over because my low back and hips were so tight I could not walk upright. Then I would wake up early the next morning to go for a run despite the soreness I still felt from the day before, further ignoring and aggravating the chronic pain I was in. I would come home nauseous yet pleased at having run my intended distance or time. I was also in emotional pain because of a breakup with someone in my sports community and felt isolated from a world that was both my competitive and social outlets. Worst of all, I remember feeling fat back then, despite clear evidence that I was not. How could that be? How could I be of one mind one day and another mind another day? And how could my mind ignore my body? I was confused and remained confused until a few more incidents caught my attention.

Not long after seeing the Facebook photo I had another similar experience. I had been in Mexico for my friend's stagette. Six of us were staying at a beach house, and we spent most of the time in our bathing suits. I wore a bikini and I remember feeling pretty good that weekend, in terms of body image. Someone took a photo and we all gathered around her digital camera to see it. I remember looking at myself confirming that I didn't look fat. "Check," I thought. I didn't feel fat and I didn't look fat. I finished the weekend feeling as good about my body as I had started it. A few months later I received digital copies of all the captured moments of the weekend, that photo among them. When the image came up, I stared at it in disbelief. I was *so* fat, I thought. I almost burst into tears in devastation. How could I have walked around looking like that and not known it! Again, I was confused. How could I be of one mind one day and a completely different mind another day, only a couple of months later?

In fact, these are not the only times in my life when this has happened. I remember playing soccer back in eighth grade. I was so self-conscious in my soccer uniform that I wouldn't run as hard as I could in order to avoid having my body jiggle in ways I wasn't comfortable with. Fifteen years later I had the opportunity to see a video of my young self again with a different mind. As I saw myself run across the field, I realized how distorted my image of my body was then compared to what felt more real in the present. I was not fat. It was an illusion of my mind. Not too unlike how I feel on this day writing, 38 weeks pregnant, looking back at photos of my pre-pregnancy self, seeing that what I thought I looked like was not at all what others would have seen.

These examples are clear and poignant to me as proof that my mind, and our minds, cannot always be trusted. I remember watching an interview with Cindy Crawford, a supermodel from the 90s, where she too said she thought she was fat, and she clearly was not. I see and hear many people thinking and feeling fat all the time, when they clearly are not. This is the basis of disorders like anorexia. Fat people in thin bodies with a mind's eye that is deceiving. But trying to tell any of us otherwise is nearly impossible until we have evidence. Eventually, the evidence comes.

The mind cannot be trusted, but knowing this fact is powerful. And being able to argue with the mind is the crux of cognitive therapy and learned optimism, which we will discuss later. In the meantime, let's explore the mind a bit more so we can prepare ourselves with more knowledge to harness its power.

The Study of the Mind

Western Psychology

The word psychology literally means "study of the mind." When Freud and Jung and other psychoanalysts were dominating psychology (late 1800s into the early-to-mid 1900s), they were primarily putting forward theories of the unconscious mind. Eventually, new theories emerged claiming that the study of the conscious and unconscious was inadequate for truly understanding the mind. Instead, bold psychologists like Watson claimed that the only things worth studying in psychology were observable and measurable behaviours. Thus, psychology became known as the study of behaviour, particularly observable behaviour. Psychologists generally began to discredit any interest in the unobservable, namely the unconsciousness activity of the mind that drove Freud and Jung beforehand. In terms of behaviour, the focus was on how our actions were reinforced, either positively or negatively, to increase the likelihood of those behaviours occurring in the future. Pavlov and his dog best exemplify this. Pavlov's dog learned that when a bell rang he would soon get his dinner. Pavlov was actually studying digestion, but with keen observation skills, he noticed that the dog began to salivate at the sound of the bell, not the food. This made no sense, biologically, because the bell was not considered part of eating. Only, it *had* become part of the eating ritual. It became a signal that the dog associated with food. As a result, the dog learned new conditions under which he would get to eat, and those conditions became known as "classical conditioning." This gave rise to a whole new field of understanding how behaviours (and we as humans) are shaped, two ways being reinforcement and punishment. Punishment can be powerful in creating aversions to certain experiences. It creates much fear, but generally speaking, it is not a good deterrent for preventing behaviour in the future.

Later, other psychologists became interested in "thinking" and thus, the emphasis on cognitive psychology emerged. Cognitive psychologists felt that focusing entirely on reinforced behaviours and punishments did not help our understanding of more complex aspects of ourselves like language, information processing, decision-making, rationality versus irrationality, and so on, especially when the complexities were beyond simple reward and punishment models. Cognitive psychologists believed in the power of the mind, similar to Freud and Jung, but developed ways to test their theories. This branch of psychology continues to be a major field of psychology and neuroscience as is behaviourism.

Today, the word psychology (according to Western psychologists) has numerous definitions incorporating many of those early elements. My definition of psychology is that it is the study of the mind and its functions; for example, thoughts (cognition), actions (behaviours), and emotions (affect). I also think that the study of the mind requires the study of not just one mind in isolation, but the relationships among minds, which is what social psychologists focus on in their studies.

The study of the mind and behaviour in psychology has come to include an exploration of the neurobiology that gives rise to the functions of the mind. Twenty years ago this may not have been the accepted reality, but today it is. Twenty-two years ago (in 1993) I was starting my undergraduate degree in psychology in Winnipeg and was about to take my first biological psychology course. A few students in the university were doing a double major in psychology and biology, an older version of a neuroscience degree that is now widely available. However, at that time there were no undergraduate neuroscience degrees in Canada. But the 90s was deemed the "decade of the brain" and, indeed, it proved to be an incredible era of discovery. By 1998, when I attended Brock University, Brock was being granted permission to launch the first undergraduate neuroscience degree program in Canada. Now, many universities have such programs, and for good reason. Many young people are interested in the biology of the brain. Together, the elements (cognition, behaviour, affect, social relationships, and awareness) all make up the mind and are brought to life because of our biology, in particular our neurobiology.

Eastern Psychology

The interest in consciousness and awareness is less dominant in psychology or neuroscience today, although cognitive psychologists do take some interest. In neuroscience it feels less easy to measure it, so it often falls by the wayside. Recently at the Society for Neuroscience annual meeting, I was talking with a colleague who brings neuroscientists to a monastery in India to teach monks about science — brain science in particular. When the monks ask about the brain science of consciousness, they are continually astonished when the neuroscientists have little to say about it. They simply cannot fathom how we could be studying the brain without studying consciousness. To them, it is one and the same. I agree, and I came to that conclusion myself after spending 10 days in silence observing my own mind and, as a result, my own brain in action. This silent meditation retreat was powerful beyond belief and showed me dimensions to my mind and brain that I had barely considered from my academic perspective. The experience helped solidify my interest in "neuropsychoidiology" and the personal study of one's own mind, brain, behaviour, thoughts, and emotions.

Generally speaking, Eastern philosophies of yoga and Buddhism (the ones I am most familiar with) are psychologies, with a deep focus on consciousness, awareness, and the self. They study all of this from a meta-cognitive perspective, that is, the practice of thinking about thinking. This type of thinking is unique to humans, as far as we can tell, and is likely the result of the highly developed cerebral cortex and the ways the cortex integrates with all other structures within our brains. Practices like yoga, meditation, and mindfulness are extremely helpful to understanding the mind in a way that Western psychology and neuroscience simply cannot. A deeper understanding requires self-knowledge; otherwise, we will lose sight of what we are exploring. Losing sight could be devastating, I think, because a mind as powerful as the human mind should be nurtured for good.

The mind has the power to heal us and to protect us from harm. The mind can propel us to great heights. The mind can create, innovate, and intuit. But the mind can also lie to us, destroy us, and even kill us. The mind is a powerful agent and this power should not be taken lightly. If we are to make great use of this power and allow the mind to work for us, then we are better off knowing a bit about it!

The Power of the Mind and Brain

Have you ever seen or heard about people walking over burning coals apparently unscathed? I remember thinking it was a trick, without realizing that the trickery was actually happening in the mind of the person doing the walking. I haven't walked over hot coals, but I have done things that seemed beyond my body's capabilities. At least twice during ultimate games, I jumped much higher than I expected to get a disc out of the air, one time grabbing it from a six-foot-tall guy (I'm five-three). People cheered not only because they were happy I caught the disc, but also because they were amazed by the height I got. I also remember having a conversation with a girl on my team once who commented on how I would just keep cutting, or running lots, in a game of ultimate, especially at times when most others were too tired to move. She asked me how I did it. I told her, "I just didn't think about any other moment except the one I was in. I pushed as hard as I could go for each cut as if it were my last one, and if another was required, I would do it again. I never reached a limit where I couldn't move so I kept cutting."

Similar examples are common among competitive athletes because we learn to keep going beyond preconceived notions of our limits. I also remember some examples from my yoga practice, which is anything but competitive. I remember when I did my first headstand, a posture that before seemed so unfathomable and inaccessible to me. But with encouragement and guidance from my teacher, who believed and had faith in my ability, it instantly became a reality. I had similar experiences in meditation. In my first *Vipassana* meditation course (described in more detail in the chapter on mindfulness meditation and yoga), we sat for 10 hours a day in meditation for 10 days in a row. For three of those hours we were to practice intense discipline and were instructed not to move. The first three days were excruciating and difficult. It felt like an endurance race. But then I wondered, what if I were making it harder than it had to be? What if I relaxed into the meditation, rather than attempting to endure it? Well, lo and behold, the moment I had that thought, something changed. I relaxed into it. I stopped making meditation so difficult. The excruciating pain I felt in my legs vanished suddenly.

Another good example of harnessing the power of the mind is in acting. Being able to express a particular emotion when one does not actually feel that emotion, requires a great deal of mind over matter. I learned that first-hand as a strategy for getting out of trouble when I was a kid. I often did things that I thought were fine (like sneaking out at night or drinking underage), but that my parents would not have approved of. When I got caught, I learned to appear very disappointed in myself and get really upset in order to lessen the punishment. If I showed enough remorse, I could get out of punishment altogether. But more relevant to the topic of this chapter was *how* I did that. I knew I could make myself cry instantly if I thought about my dog dying. To this day, I can still send myself into a downward spiral with thoughts of my dog or cat or loved one dying, but I don't do it to get out of trouble anymore.

Mind-Body Connection Exercise

Explore how your body responds to your thoughts with the following exercises. Do each one for about a minute and elaborate as much as you can. Bring as many details as possible into your mind. During and afterwards take notice of how you feel in your body. Consider physical sensations in your chest, shoulders, head, stomach, skin, or anywhere else that sensations arise.

Think about something that makes you sad. Let it resonate inside your mind.

Think about something that makes you angry. Let it resonate inside your mind.

Think about something that makes you happy. Let it resonate inside your mind.

Think about something that makes you feel loved. Let it resonate inside your mind.

Notes:

In the exercise above, we experienced how our minds can change and give rise to many different sensations even when we and our physical bodies have not moved an inch. In this situation, we are using the power of the mind to soar through time and experience. We can also use this power to soar into the future and experience the wealth of emotions that exist as possibilities. Typically, that makes us feel good. The flipside is that we can just as easily allow the mind to focus on experiences that make us feel bad, unwell, or even stressed. As in the exercise above, we likely felt those emotions of anger and sadness just by calling up thoughts of them in our minds. Although we chose to do that consciously for the sake of understanding, we often do that unconsciously, and it can affect our mental state and corresponding body experiences. That's what's going on when we let one stressor trigger a cascade of stressful thoughts, sometimes at 4 a.m. when we are trying to sleep. That's the power of the mind.

A Brain that Heals Itself

One of the clearest examples of how powerful the mind truly is — and one of my favourites! — is the placebo effect. This common effect presents as a nuisance to scientists and medical practitioners, and occurs when people are told that they're being given something with curative or healing properties when in fact what they receive has no known active healing or curing ingredients. It might be a sugar pill, for example, and it is considered an inert substance according to pharmacologists (people who study drug action). The interesting part of the placebo is that even without a healing agent, the "inert" substance might be healing if the person taking it is led to believe that the substance does, in fact, have healing properties. Imagine, for example, that you were stuffed up with a cold. If you were given a glass of lemon water but told that it contained a decongestant, you might very well experience easier breathing associated with the belief of having just taken a decongestant. That phenomenon is the placebo effect, and it represents the power of thinking, the power of our belief, and the power of the mind.

The (Neuro)biology of Belief

The placebo effect is well documented in science and medicine. Examples of observable bodily changes triggered by belief in (but absence of) any actual medical treatment include: heart rate, blood pressure, coronary diameter, gastric motility, bowel motility, lung function, chronic fatigue, irritable bowel syndrome, asthma, immune responses, pain, and itching. Epilepsy (reduction in seizures) and reduction in symptoms of depression are also known to be subject to placebo effects. This list shows how belief and expectations can alter bodily conditions from the very basic (heart rate) to the high-level (pain, immunity, and epilepsy). Not surprisingly, significant interest has been generated in the placebo effect over the past few decades. In 2005, neuroscientists Benedetti, Mayberg, Wager, Stohler, and Zubieta claim that public interest is growing in the placebo effect "because [placebo effects] promise increased self-control."[16] They also suggest

16. Fabrizio Benedetti, Helen S. Mayberg, Tor D. Wager, Christian S. Stohler, and Jon-Kar Zubieta, "Neurobiological Mechanisms of the Placebo Effect," *The Journal of Neuroscience* 25, no.45 (2005): 10390–10402.

that the "existence of placebo effects suggests that we must broaden our conception of the limits of endogenous human capability." Endogenous simply means "generated from within" and speaks to a very important point about our own inner healing capacity. Although this concept sounds cliché to some, the placebo effect and any corresponding neurobiological evidence would suggest that, indeed, we do have an inner ability for self-healing. And this capability should not be underestimated, but explored more and harnessed. Benedetti and colleagues do believe there is significant interest from the scientists' perspective because "the effects of belief on human experience and behavior provide an entry point for studying internal control of affective, sensory, and peripheral processes. The study of the placebo effect, at its core, is the study of how the context of beliefs and values shape brain processes related to perception and emotion and, ultimately, mental and physical health." Essentially, the neuroscientists would also agree that the power of belief is a brain-based phenomenon.

As a result of realizing this incredible phenomenon, the placebo effect has gone from a scientific nuisance to a neurobiological curiosity with treatment potential, at least for some neuroscientists. That said, there is difficulty in clearly outlining which areas of the nervous system are involved because the placebo effect depends on the type of placebo manifesting itself. For example, when a placebo effect is manifesting in pain reduction (analgesia), there is a change in the pain-related neurotransmitter system (the opioid system), but when the placebo effect manifests in immune responses, specific immune chemicals are altered. The placebo effect does not appear to be restricted to one region of the brain, but likely requires the participation of several areas depending on the belief and the symptoms expected to be relieved. One system that does have a rather clear role is the brain's reward system — not too surprising because relief of pain is rewarding. The brain's reward system communicates largely by the neurochemical dopamine. One of the major brain structures involved in its communication is the nucleus accumbens. In the lab, this structure is involved in a lot of addiction research because stimulating it, electrically or with drugs, is highly rewarding. If animals are left on their own with free access to the stimulator themselves, they will spend most of their day just sitting around self-stimulating. It's that rewarding! Other brain areas suspected to participate in the placebo's pain-reduction effects include areas of the cortex, specifically the dorsolateral prefrontal cortex and the insular cortex.

Despite knowing the exact nature of the placebo effect, it is clear that it is anything but benign. There are clear and well-documented changes in experience that accompany the placebo effect, and from a practical purpose, this gives much hope to those of us interested in harnessing the power of the mind (or the power of expectations). It reminds me of the phrase "fake it till you make it." In many ways, that's the placebo effect in action. If I am led to believe that I will heal, I can heal. If I am led to believe that I will be cured, I can be cured. If I am led to believe that I am a smart and talented athletic star… perhaps I just might become that too! This power of belief is important for giving hope, direction, and empowerment, all essential elements of personal growth, positive psychology, and life-coaching. More important, it shows that the power of the mind can also be explained through actions within the brain, making it a tangible and biological phenomenon. For some, that knowledge itself is power in terms of taking the power of the mind seriously.

A Brain that Protects Itself

The placebo is a good example of how the mind can work for us. To illustrate another way in which the mind is powerful, let's consider situations in which the power of the mind is used to protect us, for example after traumatic experiences.

Example 1: Dissociative Identity Disorder (Multiple Personalities)

Another favourite example of mine is when our mind protects us from negative memories and traumatic experiences through a process of dissociation. In some rare, horrible cases where children have experienced severe abuse, the mind dissociates so strongly that new personalities develop. This condition is known as dissociative identity disorder (formerly multiple personality disorder). What seems to happen is that the child's mind creates new personalities to experience the abuse so that the child doesn't have to. In many ways this can be considered quite adaptive. Imagine a child undergoing such extreme abuse but being unable to escape. The child's main source of care may be from the abuser him or herself, which would leave the child helpless in strength and resources. Physically unable to escape, the escape comes in mental form.

Example 2: Recovered Memories

A similar but less severe reaction is amnesia — a loss of the memory of any traumatic experience. I recently had a client who recovered a memory (albeit, not through our work together) of the childhood abuse that she suffered at the hands of her primary caregiver, a caregiver with whom she still had a very strong relationship. The intense memory invaded her mind one day after a lifetime of not recalling *any* such abuse, at least not consciously. These types of memories are repressed similarly to the way dissociative identity disorder works, but the memories may emerge when the individual is equipped (cognitively, socially, emotionally) to deal with what they could not have dealt with in childhood. In childhood, dealing with such trauma means repressing the experiences in some way or another. But it's interesting that these memories do not appear to be gone forever; they simply lie beneath the surface of awareness, protected by the brain's ability to be conscious and unconscious.

Example 3: Alcohol-Related Damage

A third example of how the mind attempts to protect us is not so much after a trauma, but in attempts to ward off a trauma to the ego when we are missing details to make sense of our experiences. In a neurological condition known as Wernicke-Korsakoff syndrome, individuals suffer damage to their brains due to excess drinking of alcohol, likely at least partially because of thiamin (vitamin B1) deficiency. The damage appears most in subcortical brain structures, known as the mammillary bodies in the midbrain, although other brain areas also appear to be affected, including the thalamus, hippocampus, frontal lobes, and cerebellum. These structures are crucial for memory formation and the high-level thinking known as executive functions. When these functions fail, the person loses important details necessary for memory. The brain's strategy to deal with this memory loss seems to be to fill in the missing pieces creating a seemingly logical story "confabulation", which is not simply lying. Rather, it seems to be a best guess in order to protect the individual from his or her own fragmented memory. It's as though the mind is attempting to protect the person from the reality of the unfortunate and confusing situation. Confabulation may involve prefrontal cortical regions such as the ventromedial and orbitofrontal areas and might also occur in individuals with other dementias.

Those are just some of my favourite examples used to highlight the protective power of the mind. In each of these cases though, some sort of disordered state either emerges as a result or is inherent in the manifestation of the protection. These examples show how the power of the mind can go awry. To fully use the power of the mind, we must understand the depths to which it can take us.

A Brain that Harms Itself

Example 1: Nocebo Effect

Earlier we saw what can happen when we believe we are given a healing agent. An opposite phenomenon can also occur if we believe we are taking a substance that will make us sick. This is known as the "nocebo effect." Believing that a substance will cause an effect can lead to that effect emerging, whether it's positive (placebo) or negative (nocebo).

When a nocebo effect is present, it's because we have come to believe that something will be harmful, make us sick, or cause pain, for example.

In one recent study,[17] researchers were examining this very principle with cream placed on each of the participant's arms. The participant was told that the cream on one arm was the control cream and on the other arm was a cream that would increase pain. Indeed, the nocebo effect emerged. The participant reported more pain on the arm in which the "painful" cream was placed. Although this may seem like it's entirely in one's mind, what happens in the mind is represented in the brain. In fact, we can actually trace the neurobiological response to the nocebo effect. For example, the pain signal travelling through the nerve from the nocebo arm was more strongly activated compared to the nerve in the control arm. So although there may be a subjective rating of the pain, the participant indeed registers the pain as stronger on one side. And that pain registers all the way up through the spinal cord and into the brain, giving rise to the neurobiological power of belief.

Another example of a belief that can harm has been described by health psychologist Kelly McGonigal,[18] in a summary of compelling research showing that the harmful effects of stress are only harmful if we believe them to be! When I first heard this, I was in a state of disbelief, as was McGonigal (so she admitted). She had long preached the

---

17. Stephan Geuter and Christian Büchel, "Facilitation of Pain in the Human Spinal Cord by Nocebo Treatment," *Journal of Neuroscience* 33, no. 34 (2013): 13784–90.

18. Kelly McGonigal, "How to Make Stress Your Friend," *TED Talk* (December 21, 2010), www.ted.com/talks/kelly_mcgonigal_how_to_make_stress_your_friend.html.

typical message that stress is harmful (as have I!), yet the scientific evidence suggested otherwise. Only those who believed that stress was harmful were experiencing the harmful effects, generally speaking. It's a simple and great example of a nocebo effect!

In many ways this is classic psychology that focuses on the negative, the abnormal, and the dysfunctional. We can easily turn a nocebo into a placebo with an easy change to our language. We'll explore the practice of how we word things in the chapter on positive thinking and optimism.

The nocebo is also a good example of how the mind can turn against us. So are the examples of when the brain attempts to protect. In early phases, the attempts to protect appear adaptive, but in later phases they become maladaptive, kind of like a light switch that never shuts off. During the day it's helpful, but when we are trying to sleep, it's not so helpful. There are other conditions in which a normally adaptive system goes awry.

Example 2: Anxiety

Anxiety is a great example of the mind's power gone awry. Anxiety is a form of fear, albeit an irrational, unfounded, misdirected, or maladaptive fear. Fear is normally part of the fight-or-flight stress response, which is an acute, adaptive physiological system that allows us to move ourselves out of danger's way quickly. A variety of processes occur in the body to prepare us for fight or flight, which will be described in greater detail in the chapter on stress and self care. Briefly, in a normal fight-or-flight response, the individual and his or her body respond to a real-life threat of danger. When we hear heroic stories of people using ridiculous amounts of strength to lift a car off a baby or carry an adult out of a burning house with apparent ease, we are privy to the power and physiology of this fight-or-flight system. The mind, perceiving something terribly dangerous, activates the body to do incredible feats.

This stress system evolved to serve in those rare moments when life is being threatened. But when this powerful system is used in perceived (rather than real) circumstances, it causes havoc. It's maladaptive to use precious resources to fear, say, a conversation with your boss, an upcoming deadline, or a public speaking event — all events that are, generally speaking, not actually life-threatening. Many of us react as if these events are life-threatening simply because somewhere in the mind (and brain) we have perceived something wrong and overreacted. In so doing we activate a system that was not designed to be used in that way. When we perceive a true threat, it's great to have the system activate. But when we sound the alarm bells too early, it's the result of an overly sensitive stress system and a mind too reactive for its own good.

Example 3: Depression and Suicide

Similar to anxiety, depression can also be attributed to misperceptions. These misperceptions show up in the way we explain things that happen to us, particularly bad things. Explaining things in a pessimistic way is a risk factor for depression. Positive psychologist Martin Seligman describes this process at length in his book *Learned Optimism* and we will explore it further in the chapter on positive thinking and optimism.

Related to depression, suicide is one of the most convincing examples of the power of the mind. Suicide is a devastating human experience for those who battle with the idea of it internally and for those who have lost a loved one to it. The body's sense of physical survival is trumped by the mind's need for pain relief and escape. The mind, in this situation, is much more powerful than the innate physical need to survive. I think of a friend of mine, who was taken to the hospital twice, once by me, after she took a bottle of pills. The pain was too great, she said. She wanted an out. She didn't think she could survive this life. She heard the voice inside pleading for some kind of relief, some kind of escape from the pain that she experiences. But for her, the thoughts served as a message that something needed to change after pressure built up and consumed her mind. Her thoughts were a warning sign. People, or their minds or their brains, choose suicide because it feels like the only option. If that's not the best example of how powerful the mind is, I'm not sure what is. But if we could harness that power and channel it in another direction, I suspect we could move mountains. For example, I have no doubt in the power of my friend's mind. She is an athlete and has demonstrated many times her mental strength in both hockey and running (training for and completing marathons twice). She is smart and has done well at school. She is a doctor and runs her own business very successfully. Most important, she has this incredible power to hold other people's pain, inviting them to heal through vulnerability. She is loved by many and has brought many people and communities together. Her power spans far and wide, often because of her openness with experiences like the one's described here.

I know many people affected by suicidal ideation, all of whom have powerful minds. But their power must be taken seriously and not misdirected back onto themselves with negativity. Another friend comes to mind. He has been suffering with depression since he was young. As a kid he used to complain about feeling sick, but his doctors never found anything physically wrong with him. Today, we know those symptoms often indicate depression in young people. At the time he was experiencing them, they were simply "in his head," or so he was told. But in reality he was reacting to something. I have seen his mind at work. He is smart and easily grasps concepts. He is also filled with creative ideas, both business and research ideas. But somewhere along the way he picked up the belief that he wasn't good enough or that he would fail, a common misperception of people with depression. The fear of failure becomes too great for the individuals to continue forward.

Another person I know experienced a string of other people's deaths by suicide as a child. She grew up with her own suicidal thoughts and thought that everyone had them. In her life, they were normal. For her, suicide was a normal way of dying and thinking about taking her own life was equally normal. She owned those thoughts as not only her own, but as a normal way of progressing through life. Not surprisingly, she still has these thoughts today. She has made numerous attempts at her own life, and she has been in and out of treatment programs. I know she has a powerful mind. I have seen her way of thinking, interpreting, and understanding other minds. But I'm not sure she realizes how beautiful and powerful her mind is.

Harnessing the Power of Mind and Brain

The mind is powerful. On the one hand it can heal and protect itself, but on the other hand it can wreak havoc and damage itself. *Quod me nutrit, me destruit* — that which nourishes me, destroys me. This quote resonates with me deeply because in many ways I feel both extremes. I am at the whim of my mind, and the only thing that saves me is a long history attempting to know and understand it. It heals and nourishes me. As far as I can tell, I have been able to ward off colds, stop a burn on my finger, transform pain into sensation, see the silver lining in terrible situations, empathize, offer compassion, play harder, longer, and better in sports, and find energy to overcome exhaustion when needed. My mind also entertains me, is creative, thinks deeply and holistically, forms opinions, and focuses for long periods of time so I can complete the work I need to get done. I have been driven and determined, and I have accomplished a lot because my mind made it possible. But my mind also destroys me. I have had terrible days when I thought my body looked awful, and I could crumble to my knees with self-hatred. I have felt stupider than I ever thought possible. I have lost purpose, succumbed to failure, and been paranoid to the point of panic. And my worries about death and dying have caused me to wake up in the middle of the night with my heart racing.

I have witnessed my mind fluctuations and its delusions and when it goes awry. When I was young and I began staying at home by myself while my dad was away, I would go around and check every door and window twice before going to bed. I would also tie pots and pans to rope, stringing them up on the door so that if the door opened, they would come crashing down on another set of pots and pans, all of which would wake me up and give me enough time to react. I would sleep with my cordless phone under my pillow so I could quickly call for help. And I would have my hockey stick in bed with me as my weapon.

At times, I have wondered about my sanity (or lack thereof). I have worried that I would lose my mind or had already done so. I know parts of my mind cannot always be trusted. But I also know that I can direct my brain for good. I can see the good, the beautiful, the excitement of the life I have. I can see that I have accomplished a lot of things in my life, thanks to my mind. My mind has created communities, scientific knowledge, sports teams, leagues, and new divisions at national championships. My mind has seen and created equal opportunities for women. My mind has created my own life-coaching practice out of nothing, a philosophical approach to accompany it, and then a certification course to train others. My mind has created a company and a non-profit organization. My mind is idealistic, innovative, happy, and full of love and humanity. Knowing that my mind could harm itself is worrisome, sure. Knowing my mind can protect itself is comforting. But knowing that my mind could heal itself and create a wonderful existence is something extraordinary. That existence is worth achieving.

Harnessing the power of the mind is simple, in principle, but it requires practice that involves a few basic elements:

- Explore your brain (with or without fear).
- Know your brain (through exploration and discovery).
- Channel your brain (with intention and diligence).

That's all we really need to know, at a basic level. Sure, there are tools that encourage each of those, but generally it's just practice, intention, and direction that gets us where we want to be.

Chapter Assignment:
Self-Reflection Questions

Quod me nutrit, me destruit. That which nourishes me, destroys me. My love for learning nourishes me but it also destroys me. I can get so caught up in my own mind and thoughts that I forget almost entirely about my body. When I was doing my PhD, for a few weeks I would completely ignore the signs that I had to go to the bathroom because my office was about a one-minute walk away from the nearest washroom. I would be so occupied with my mind's work that I just would not go. I remember a few times it got so bad that I would find myself running to the bathroom, barely making it. Then one time ... I actually peed my pants! My mind was so caught up with learning and studying and thinking (my nourishment) that it essentially destroyed me ... or at least my pants!

Have you ever had an experience where that which nourishes you has destroyed you?

How else does your mind harm you?

What beliefs or expectations are holding you back?

What beliefs or expectations are supporting you to go further and deeper?

How does your mind protect you?

How does your mind protect you in a way that no longer serves you?

How does your mind nourish you?

How does your mind heal you?

How does your mind support you?

How does your mind empower you, give you strength, and lift you beyond your expectations?

How *can* your mind empower you more, give you more strength, and lift you further, beyond your expectations?

Notes:

Section II

Thoughts, Actions, and Emotions

Chapter 4

Stress Reduction and Self Care

It's almost cliché to say we all experience stress, but the truth is that we all should. Stress is a good and natural part of our existence. YES. It is! Stress is an adaptive biological response used to mobilize the body and divert resources away from secondary bodily functions (like digestion) toward those useful for saving us from danger (like increased blood circulation to get oxygen to muscles that are needed to help us move and escape). Using very precise physiological changes, the body, mind, and brain kick into gear to prepare us for fight or flight. Being able to quickly divert resources away from non-essential activity so we can avoid life-threatening situations is very adaptive. Imagine what would happen if you were faced with a real-life threat, be it a bear in the woods, a bus that was about to crush you, or an attacker in a dark alley. Wouldn't you want the capacity to act quickly and move out of the way of that danger as soon as possible? That is what the stress response is for.

Physical and Psychological Stress

The distinction between physical stress and stress in the mind is not very helpful to understanding the nature of stress. It misconstrues psychological stress as being entirely in the mind when, in fact, stress is stress and the end result is essentially the same. The fact that we can conjure up reasons to be stressed is a separate issue, an issue concerning the brain. It is fair to say that some things that tax the body are physical stressors. An infection in the body causes physical stress. Sleep deprivation causes physical stress. Using muscles to the point of exhaustion causes physical stress. Marathon running causes physical stress. Birth causes physical stress. These events can cause mental stress too, but even without the mental aspect, they tax our bodies. But so do deadlines, interpersonal conflict, weddings, changes in jobs, buying a new house, and worrying about money. These latter factors are psychological merely because we have the ability to conjure them up in our brains as threatening to us. But all of these stressors signal to the body that danger is present and result in a cascade of physical events within the body and brain. A problem still arises when we imagine dangerous situations in the mind even when the situation is not life-threatening. Think of the word "deadline" for example. Or the phrase, "I'm drowning in work" or "I'm suffocating in my relationship." These are psychological stressors. Psychological stress is a peril of our perception and complex thinking resulting from our incredibly well-developed frontal lobes. The frontal lobe has the capacity to imagine a different future, to overanalyze, and to ruminate. These psychological experiences essentially become physical realities in our body, another example of the power of the mind.

Physical Changes Associated with the Stress Response

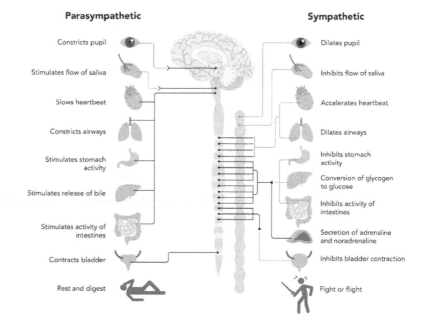

Figure 4. Sympathetic and Parasympathetic Nervous System

Exercise

What examples can you think of from your own life that give rise to some of the physical experiences described above? List them.

Notes:

Consider these two common stressors:

Public Speaking

The fear of public speaking is common to many people. They start to sweat, get shaky, maybe develop a funny feeling in the pit of their stomachs, or nausea, or clammy hands. Their throats might close up and their mouths might dry up, and they may also experience muscle tension. All of these, as we will see later, are physiological responses to stress. But think about it. I just said that stress is a very natural adaptive system designed to move us quickly out of the way of a life-threatening danger. With something like public speaking, we really aren't in a physical situation. Sure, we may fail, look foolish, and have our egos severely bruised if we screw up, but we shouldn't die because of it. But the worry of how horrible it would be and feeling like we will "die up there," or the thought that public speaking would kill you exemplifies the fact that sometimes we create life-threatening situations in the mind, situations that don't really exist in any physical way.

Traffic

I grew up in Winnipeg, where driving around the city is a pretty awful experience when you're trying to get somewhere during rush hour. The worst time is during the short summer season when the city races to complete road construction before the next winter hits and the roads freeze. The consequence is significant time added to the daily commute. But of course, this is after a long winter when people have to drive cautiously because of persistently slippery roads. Needless to say, driving can be stressful for many people. We sit in our cars, scratching our heads to alleviate some of the building, pent-up energy. What is that energy? It's the fight-or-flight system. Symptoms include racing heart, clenched fists and teeth, and a headache, to list a few. This is our body ready for a fight. And we do fight for our space and fight to be quicker. That's where road rage comes in. Yelling, screaming at others inside the cars, rolling down windows to be more convincing, adding in a finger here and there, and cursing and swearing. And of course, *never* letting anyone in ahead of us. We are fighting for limited resources here! But again, what are we actually preparing ourselves for? Is this fight necessary? Would fleeing be better? It would probably serve us better to realize that we are not actually in a life-threatening situation. We will not die if we are late for work, for our appointments, or for getting home simply to watch TV. So why bother engaging our stress systems? Because our minds can!

These examples highlight how our mind can cause havoc with our system, how it can turn on itself and cause real physical problems. Indeed, that is what happens under psychological stress. A common misunderstanding about psychological stress is that it is all in the mind and, by extension, that it isn't all that bad. Misinformed people sometimes believe that psychological stress is a sign of weakness, even a personality flaw, and they may not take it seriously. But in fact, psychological stress is just as problematic as physical stress because it does actually lead to physical changes in the body. Most of us are actually experiencing psychological stress when we experience stress because most of us are not out in the woods, running for our lives or trying to redirect resources because we have a real-life predator after us. We are not usually taking steps into oncoming traffic and having to jump backwards to avoid being crushed. Our stress is primarily psychological because it only exists due to perceptions we hold. Because so much of our stress is in the mind, so to speak, programs like mindfulness-based stress reduction have proven very effective for reducing stress and even healing issues in our body, such as psoriasis.

The Neurobiology of Stress Hormone

When we perceive a stressor in our environment, several things happen in the body, through the various divisions of the nervous system. Remember that the nervous system is divided into central and peripheral nervous systems, and the brain and spinal cord are part of the central system. The peripheral nervous system is associated with everything peripheral to the brain and spinal cord. It includes another division, called the autonomic nervous system, which controls a lot of the automatic functions in the body that we don't think about. Arousal and response to a threat (or stressor) is one of the things we don't have to think about. The major pathway that invokes the stress response is called the "sympathetic nervous system." Activating the sympathetic nervous system leads to a whole host of physiological events and body changes, including increased heart rate, increased blood flow to the muscles, sweating, changes in eyesight and brain awareness to increase vigilance, and increased activity of the adrenal glands to pump out stress hormones (for example, adrenaline, or epinephrine). It also leads to a suppression of digestion through decreased salivation (dry mouth) and fewer resources being directed to the gut and reproductive organs. This prevents energy being wasted on breaking down foods when the body needs the energy for getting out of danger. Recently, I made my students deliver presentations in class. Sometime later I was teaching about stress and one student asked if all of those feelings she was experiencing during her presentation were representative of the sympathetic nervous system. Indeed, they were!

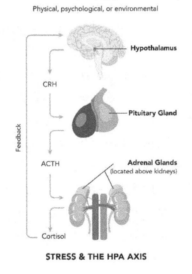

Figure 5. The (H)ypothalamic-(P)ituitary-(A)drenal Axis

Another part of the nervous system that is involved in the stress response begins with the activation of the brain structure called the "hypothalamus," which is an area of the brain that also regulates things like feeding, drinking, sexual behaviour, and stress. This pathway ultimately leads to elevated levels of the stress hormone. In order to raise these levels, the hypothalamus sends chemicals (hormones) in the brain to target the pituitary gland, which dangles right underneath the hypothalamus. The pituitary releases another hormone directly into the blood stream, which circulates through our body until it reaches the adrenal glands that sit near our kidneys, behind our hip bones. The adrenal glands also release a hormone called "cortisol" into circulation, and that hormone is known as the "stress hormone." You may be familiar with the word "cortisol" because, under several different names (cortisone, hydrocortisone, prednisone), it is the basis for some common anti-inflammatory medications that are used to treat conditions like eczema, poison ivy, and even painful joints in many athletes.

Cortisol is a very powerful steroid hormone. Much the way neurochemicals are derived from building blocks in our diet, so is cortisol. Cortisol is derived from cholesterol. Etymologically, "sterol" indicates its relationship to steroids. The other steroid hormones that you are more familiar with are sex hormones such as androgens (like testosterone) and estrogens (like estradiol). Testosterone is also derived from cholesterol. The word testosterone comes from the combination of the words "testicle" and "steroid." Estrogen is derived from the conversion of testosterone and is therefore also a steroid hormone.

Hormones do more than just alter moods and drive sex. Hormones are fast-acting chemical messengers that circulate through the bloodstream and serve as a communication tool among different organs, forming the endocrine system. Hormones are produced selectively within major endocrine glands, such as the pituitary glands. Others you may know of include the thyroid and thymus glands in the upper chest and throat region; the adrenals and the pancreas near the kidneys; and the gonads (ovaries and testes in females and males, respectively). Hormones are carefully regulated in the body to keep us in balance.

Cortisol Effects in the Brain

Cortisol (and the other steroid hormones) exert their effects in several different locations by travelling through the bloodstream to arrive at their target organs. The brain is one of those target organs. One area of the brain with a lot of receptors for cortisol (and estrogen) is the hippocampus. The hippocampus is involved in memory and place, where we code information for where we are. It is helpful in changing our environments from feeling novel to feeling familiar. Because the hippocampus has a high density of receptors that respond to cortisol (known as glucocorticoids), when our HPA axis is activated and cortisol is released from the adrenal glands, our hippocampus receives that message loud and clear. It receives a message that something is wrong, that danger is present, and uses the information it gathers, along with information from another structure, the amygdala. The amygdala is a central structure involved in processing the emotions associated with stress and anxiety. Together, the amygdala and the hippocampus work to either keep the HPA axis activated with continued threat or slow it down when the threat dissipates. The amygdala will keep the circuit going. The hippocampus will put on the brakes when, after properly surveying the context, situation, and the environment, it feels that the coast is clear.

Another way to understand the role of the hippocampus and the amygdala is to consider this scenario. If all of a sudden you saw a bear during a walk in the woods, I could tell you that your amygdala would be working hard with your hippocampus to alert you that there was danger in your path. The amygdala is involved in processing (and remembering) the emotional experience; whereas the hippocampus is involved in processing the cues in the environment and assessing an escape route. Both of those structures are part of the "limbic system" of the brain, a group of structures that form an emotional "network" in the brain.

Under normal situations, this system works very well for its purpose. The hormones and brain structures activated in stressful situations can even enhance memory and increase physical performance. Using this wisely and efficiently, this arousal is the nature of performance management done with elite athletes, musicians, actors, and speakers, for example. A good level of arousal is adaptive both in terms of getting us out of danger and letting us shine. Learning to use this wisely, rather than letting it run away with the mind, is also highly adaptive. Too much stress on the other hand, damages the brain and interferes with the body's ability to use the stress response wisely in the future.

Stress Damages the Brain

The adrenal glands participate, along with the pituitary gland, in the regulation of the stress response. They produce sex hormones and the well-known hormone epinephrine (or adrenaline). These glands are activated every time we encounter a stressful event, so you can imagine how they get overused when there is too much stress in our lives. In fact, because these glands are part of a loop that involves the pituitary and another region of the brain known as the hypothalamus, repeated stressful events can actually cause brain damage.

One problem with chronic stress is that the hippocampus can become worn out from the constant activation of its stress-hormone receptors. Chronic activity can literally damage the hippocampus. Earlier we talked about the forms of plasticity that exist in the brain. Many of those same elements are negatively affected by stress, for example, neurogenesis slows down significantly. The dendritic branching is reduced. Cells die.

Acute Versus Chronic Stress

Stressors are relatively harmless when they emerge acutely and don't last. The body's stress system is powerful, but it is meant to react quickly and then fade away quickly, restoring our homeostatic balance. Homeostasis refers to the body's natural ability to tend toward equilibrium or balance (as we more generally refer to it). Homeostasis concerns all of the biological and psychological factors that help right a person, whether a person is literally upside down, or has experienced a psychologically stressful disruption. For an organism (including humans) to find balance after these acute stressors is easy. But it's not easy to find balance with repeated, chronic, and persistent stressors.

Because the psychological stressors can result in physical changes, potential for another major problem exists with stress: chronic stress. Acute stress is good, adaptive, and necessary. There is no question about that. We need acute stress in order to mobilize ourselves and remove ourselves from danger. However, chronic stress is not good and our bodies were not designed to deal with chronic stress. Consider another example of a psychological stressor. My partner launched a crowdfunding campaign the night before I wrote this. He and his business partners were under a tight deadline to get everything together for the huge campaign that was set to go live. They scrambled all week on little sleep and put in 18-hour days. This was, of course, after months of working hard to get their product ready to consider the launch. If I monitored their bodies, I might not have found all the symptoms I mentioned above. Why not? Because they were all probably way past the point of stress over-activation. They were hitting the point of chronic stress and exhaustion. Chronic stress is the other place where stress goes bad.

Chronic stress is not benign. It has several effects on the body, including one interesting phenomenon involving telomeres. Telomeres are the ends of DNA strands. Each time a cell divides, which occurs normally in the body, a piece of the telomere is broken off (or chewed up through an enzyme called telomerase), effectively shortening the DNA strand. Once enough of the end has been shortened, the cell dies. This process can be quick or slow. Ideally we want the process to be slow to preserve the life of our cells (and by extension, ourselves!). But under stressful conditions this telomerase activity speeds up, and our cells end up dying more quickly. The result is a shorter lifespan for our cells and, therefore, ourselves.

Hans Selye was one of the original scientists to theorize about stress, adaptation, and exhaustion with his model, called the general adaptation syndrome. According to his model, stress involves three stages: 1) alarm reaction, 2) resistance, and 3) exhaustion. When we coexist normally with stress, we experience alarm reactions, but then they subside, and we go back to our business. But under chronic stress conditions we are forced to live in a state of resistance, and if that persists, we enter into exhaustion. Although we have incredible capacity to cope and be adaptable to our circumstances, our "adaptation energy" is finite. Indeed, there are many stressors in our life that threaten our adaptive energy. "Anything that causes stress endangers life, unless it is met by adequate adaptive responses; conversely, anything that endangers life causes stress and adaptive responses,"[19] Hans Selye stated in the *British Journal of Medicine* in 1950.

Relaxation, Rejuvenation, and Stress-Reduction as Basic Forms of Self Care

The body is designed to work in balance. So the sympathetic nervous system has a counteracting system that serves to calm all of those activities back down, called the parasympathetic nervous system (also known as the "rest and digest" system). When the sympathetic nervous system is aroused, the parasympathetic nervous system eventually brings us back down. However, with chronic sustained activation of the sympathetic nervous system,

---

19. Hans Selye, "Stress and the General Adaptation Syndrome," *British Medical Journal* (June 17, 1950): 1383–1392, http://europepmc.org/backend/ptpmcrender.fcgi?accid=PMC2038162&blobtype=pdf

the efforts to balance get worn out. The parasympathetic nervous system fails to respond and doesn't do its job very well. The sympathetic nervous system also becomes sensitized and can more easily be activated.

We need to let our bodies relax, and this isn't just some fluffy feel-good prescription. It's imperative to the entire bodily system that we have time to relax our nervous systems. As mentioned, the stress system is not designed to work on overdrive continuously. It is designed to be short-acting. When forced to work chronically, problems arise.

We need to let the sympathetic nervous system be quiet so we can rejuvenate our body. Energy needs to be redirected back to our other important organs and functions like digestion, reproduction, cellular repair, and sleep. Our body needs to relax so we can focus on those other functions that are important for long-term survival. Those functions were temporarily ignored because if we don't survive the moment of danger, then digesting our food, ovulating, and fighting off infections don't matter. But we need the moments of arousal to subside so the body can get back to regular operation. We need to reduce our stress and we need to maintain a health and self-care plan.

Stress reduction and self-care plans are not selfish practices. They are necessary in order for our own survival and, at times, necessary for the survival of those who depend on us. If we can't take care of ourselves, how can expect to take care of anyone else, be they clients, friends in need, or family members. Caring for ourselves should be a top priority if we expect to contribute positively to those around us and to the world at large.

You can rack your brain with justifications of why you are not important, but I am telling you (and your brain) point-blank that you are, and if you don't take your own care seriously, others will have to pick up the burden.

Stress, Epigenetics, and Plasticity

Stress also exerts itself through epigenetic changes within the brain. Earlier we were introduced to the concept of epigenetics and the nature-nurture debate. Epigenetics is the study of how our DNA (and genes) are modified through the tightening and relaxing of the DNA strands. These changes are interesting because they can also be passed on to the next generation, providing a biological mechanism for a parent to pass on life experience, including stress!

Canadian scientist Michael Meaney has been researching this phenomenon in his lab. Meaney has been using a rat maternal care model to investigate. He has shown that when rat pups receive good maternal care (for example, lots of licking, grooming, and nursing) they grow up to be less fearful, show fewer signs of physiological stress, and, if female, provide better care for their young. He has also shown that these behaviours run in families. Having a good mom means that a rat is more likely to be a good mom herself, and also means that her offspring will be less fearful, show fewer signs of stress, and also become good moms.

These behaviours continue across generations unless of course, you're swapped at birth! Meaney placed pups born to bad moms with good moms and vice versa. Surprisingly, the pups took on the behaviours associated with the maternal care they received, not the behaviours associated with the maternal care of their biological mothers. What this meant was that the environment had a greater impact on how rats would turn out as adults than did the genetic lineage.

These results were so fascinating that they were published in the very prestigious journal *Science* in 1999. In 2004 Meaney's group showed that these effects were happening *epigenetically* and were passed on to subsequent generations. They found changes in DNA methylation and histone deacetylation that were associated with both the rats' early life experiences and the behaviours and physiology they developed in adulthood. And even though the changes were passed on, if the pups were swapped at birth, the effects were reversed, further confirming they were epigenetic and not just genetic.

Even more convincing evidence of an epigenetic mode of inheritance was uncovered when Meaney's group chemically blocked the DNA methylation and histone deacetylation in the pups raised by bad moms. When they blocked these epigenetic changes, the pups grew up to behave and show stress behaviours like those who had been reared by good moms! This shows, without a doubt, that epigenetic changes were necessary for the early life experiences to dictate the future of the pups. And when those changes were blocked, the environment could not leave its footprints.

Anxiety Is Stress that Is Not Actually Life-Threatening

Anxiety is an exaggerated and often dysfunctional form of fear. It is a fear when no real life-threatening danger is present. In many ways, it is the mind gone awry. The perceived threat, however, threatens our ego, our status, and our existence in ways that may feel like they are life-threatening, but are not in actuality. A problem with anxiety is that the fear-inducing stimulus, for example, job performance, doesn't go away quickly. It persists and causes further problems.

I have had anxiety for as long as I can remember although I didn't know it. It was just part of me, part of what I knew as me. During my undergraduate degree I began taking on way too many things, partially because there was a world available to me with so many opportunities and fun things to do. How could I possibly say no to everyone? I would immerse myself in activity after activity. I maintained that level of "to do" for quite some time, in fact I got through two degrees that way and would soon be on to another.

Coincidently, when I moved to Halifax to do my PhD, I entered into a lab where I would study the neuroscience of fear, anxiety, and stress and its development. I didn't go because I was passionate about anxiety research. In fact, I was simply passionate about all research that helped me understand behaviour. The fact that this particular lab was one of stress neuroscience was just how things worked out.

When I got to my new lab, my life sped up again! The research was new, overwhelming, and extremely laborious, and the coursework was demanding. I started drinking coffee for the very first time in order to keep up, wake up, and stay up. I was no longer sleeping as much as I needed to. I figured five hours of sleep was enough and that the extra time would allow me to keep on top of my coursework and research. On top of my academic demands, my grandmother, with whom I was very close, was dying of cancer back in Winnipeg. I was also a few months away from breaking up with my boyfriend of 12 years, who was also back in Winnipeg pressuring me to move home.

By January 2001 of that first year, I was anxious, *really* anxious. My anxiety hit an all-time high. Although my heart was pounding constantly, at first I passed it off. I thought it was "normal" stress behaviour. Indeed, it was stress behaviour. But it was getting the best of me. I could no longer go out with friends, instead preferring to come home to my empty apartment and have a glass of wine (or two or three) and watch TV. TV was my social escape, where I didn't have to do any of the socializing (because I had no energy for that) but could still feel socially fulfilled, as sad as that sounds. Eventually, I realized that my anxiety was bad enough to learn the diagnosis, and I asked my family doctor about taking anxiety medication.

I started taking my prescription for SSRI[20] (serotonin boosters) that day. I felt immediate peace, and my pounding heart moved back into the background. My body felt quiet for the first time in as long as I could remember. I felt a deep sense of calmness come over me. I felt like I was lying on a cloud looking down at a peaceful world. Cheesy, yes, but true for me. Everything seemed so wonderful! I started my prescription in January and vowed to stay on my medication till May, just to get through my first-year PhD courses. May 1st came and, as I had promised myself, I was off the meds, although, to this day, I still have the bottle of pills in my medicine cabinet.

The experience was enlightening in many ways. It provided me with the experience of calmness, something

---

20. SSRIs are described in chapter 2.

that I had believed was only available as a few isolated moments in an entire lifetime — moments like drifting away on a boat, hanging out on a dock, or just feeling the wind blow through your hair as you bike. Instances, not constants. It never occurred to me that this might be the norm for someone. It definitely never occurred to me that it could be the norm for me. It never occurred to me because my "normal" had been so different for my entire life.

That experience didn't end my anxiety but it changed it dramatically. It gave me a new baseline to seek and achieve. It gave me both a journey and a destination. In fact, I struggled with anxiety here and there afterwards. I had another episode in 2005, after I had finished my PhD. I was in a new lab as a post-doctoral fellow, and I began teaching my very first university course. My life instantly became a whirlwind. I felt like I was going crazy ... again. The term ended and I found my way out of the hole I had dug, mostly because I was familiar with that place and didn't want to stay there.

I hit that zone again in 2014. I was running my own business, which is more than a full-time job, and on top of that I had accepted two new teaching contracts at the University of Toronto, on two different campuses. I thought I was managing fine, but in reality I was not. I'd been having problems falling or staying asleep. I'd be up all night thinking about my performance and the performance of my students, and how their performance would reflect upon my performance, and then a whole network of anxiety-provoking thoughts would spin out of control as I lay there awake in their presence.

Luckily, it didn't take me 25 years to figure it out like the first time. It took me only six weeks because I knew the warning signs. I knew what anxiety felt like for me, and I knew that I was in the middle of it. Sometimes we need medication, like I did in 2001, to stop the spiral downward. Sometimes we need a meditation retreat, like the one I benefitted from in 2007, or to start a yoga practice, like I did in 2008. Sometimes we need to relax, take care of ourselves, and just rebuild our strength. Sometimes we just need to become aware of what is happening, like I finally did in 2014.

This life, and the struggles we encounter during it, provides a journey for us. I may never live a completely anxiety-free life. So the challenge (and the reward) is to find a path that meanders through this life with awareness when we fall into the zone of "too much." The takeaways from this chapter are as follows: 1) stress is generally a good thing, but should be experienced in moderation in order to keep it functioning as a good thing; 2) stress is experienced physically whether it starts with a thought or a physiological event; and 3) being able to engage a relaxation response is helpful in keeping our homeostatic balance. The biggest takeaway is that although stress is a physiological response in our body, generally speaking, we have some degree of control over it, at least once we draw our awareness to what it is, what it feels like, and how to manage it.

Self Care

The next step in developing our well-being, then, is to take action to reduce stress and, in the process, learn to take better care of ourselves. Self care is important for our health but also for the health of those around us. A great analogy for the importance of this concept is offered up when we fly. During the safety demonstration, we are told to secure our own mask before we assist someone else. If we fail to abide by that recommendation, we could lose our lifeline, and not only would we be unavailable to help someone else, but we would also risk both of us perishing. Accordingly, developing a plan to reduce stress and practise regular self care will serve the greater good.

Self care is more than just stress reduction. Self care is about learning to love, nurture, and care for ourselves in a way we might do when caring for a child. Self care involves saying "no" to some of the *shoulds* that we carry around and weigh us down while also saying "yes" more often to our own needs and our heart's desire.

Chapter Assignment Part I:
Reduce Stress

Below is a list of ways in which you can reduce stress in your life and begin to take care of yourself and your nervous system. Choose one of these suggestions and implement it as soon as possible. Then, a week later, choose another one to add to your self-care plan, then another. Do three to five if you can so that they slowly become habits in your new stress-reduced life!

> **Don't watch the news or read stimulating material right before going to bed**. That type of content arouses the stress system.

> **Plan ahead**. Use your incredible frontal lobe to avoid those terrible moments of rushing. Make lunches the night before, know your route before you leave home, give yourself an extra 15 minutes to get somewhere.

> **Put away your smartphone and computer**. Don't take your phone with you when you walk the dog, run to the grocery store, or go to the gym. If you have to have it, then put it on airplane mode. And definitely do not allow

your phone to make any sounds while you are sleeping. In fact, except for ringing, don't let it make any sounds ever. All of those sounds are incredibly distracting.

➤ In idle moments, like waiting in line or taking transit, take the time to **sit quietly and do nothing** (no phone, no reading, no talking).

➤ **Hang out with friends** in relaxing environments.

➤ **Take a vacation**. Go away somewhere exotic or do a staycation where you live. Make an active choice to take the vacation, making it different from your daily routine of working.

➤ **Practise relaxing**. Seriously, this is a practice. For many of us, relaxing hasn't been natural for a long time, so it requires retraining. Try listening to a guided relaxation, like progressive neuromuscular relaxation.

➤ **Learn to breathe deeply**. Create a habit of deep breathing. For example, put up sticky notes or set hourly chimes to remind you to take 3 deep breaths, or take 3 deep breaths every time you walk into a new room.

➤ **Meditate**. People often say they find meditating too hard because they can't focus. What they don't understand is that the thing keeping them away is the skill they will learn by practising meditation, along with a whole other wonderful set of skills. If you don't know how to meditate, set a timer for two minutes (at least), sit quietly with your eyes closed, and focus on your breath for the entire time. You will notice that your mind wanders and that's okay. Once you notice that you are distracted from your breath, come back to feeling yourself breathing, in and out, over and over again. Every time you get distracted, come back to your breath. Keep going until the timer goes off. And then do it daily!

➢ **Do yoga**. The first time I was told "yoga calms the nervous system," I thought it was just another neuro-washed marketing tactic to convince people to do yoga. But in fact, yoga does relax the nervous system. The first time I did *savasana* (the "relaxation" at the end), I was wide awake, continuously opening my eyes to see when we would be done. Then I learned to fall asleep. Now, I can lie there and actually just relax.

➢ **Work less. Play more**. If work is play for you then play less and do nothing more.

➢ **Do more of what you love**.

Notes:

Chapter Assignment Part II:
Develop a Self-Care Plan

The purpose of this assignment is to create a self-care plan. A self-care plan is twofold: one aspect is to reduce stress and the other is to increase inner peace, love, and/or joy. Although we want to reduce the negative, we also want to focus on the positive.

➢ Assess your stress. Check in with your body in response to different situations you find yourself in throughout your week. You can also fill out questionnaires online to assess yourself; the Holmes-Rahe Stress Inventory or the Smith Stress Inventory Series works well. You can also ask simple self-reflection questions like "What invokes my stress response?"

➢ Explore ways to reduce stress to help you find what works best for you. Consider doing an online progressive neuromuscular relaxation. Consider listening to a guided meditation. Enjoy a warm bath. Disconnect from the Internet. Ask others what they do. This phase may take a little trial and error!

➢ Once you have found a method you like, use it as a stress-reduction practice for the next month and commit to doing it.

➢ Assess how you feel.

➢ Treat yourself often to something (ideally healthy) that makes you feel good, happy, strong, confident, loved, peaceful, joyful, or alive. Examples might include taking a warm bath, spending time alone or with friends, going to the gym or doing yoga, or pursuing a creative interest.

➢ Learn to say no. Start by identifying scenarios where you find it difficult to say no. Then ask yourself, "What factors seem to contribute to my difficulty in saying no? What strategies have worked for me in the past in saying no? What strategies seem to work for others? What can I commit to doing in the future to help me say no more often?"

➢ Learn to say yes. Start by identifying a scenario where you find it difficult to say yes. Then ask yourself, "What factors seem to contribute to my difficulty in saying yes? What strategies have worked for me in the past in saying yes? What strategies seem to work for others? What can I commit to doing in the future to help me say yes more often?"

Notes:

Chapter 5

Positive Thinking and Optimism

How and what we think is a deeply personal matter. We take for granted that the contents of our mind are true because they seem to arise naturally. We rarely question whether that thought was real or could have been influenced by external or internal factors when, in fact, much of our thinking is not our own. What and how we think is largely a result of influences around us, whether from our teachers, parents, partners, colleagues, classmates, friends, or culture in general.

I remember realizing this clearly when I got my cat, Pickles. I grew up thinking that cats and all of their behaviours were annoying. It didn't really strike me as odd for some time that "hating cats" was quite contrary to my general love for animals. At some point when I was a child, I wanted to become a vet. I used to nurse little birds that flew into our front window or fell out of their nests back to health. I would go out looking for stray animals — everything but cats, of course. Not too long ago, I acquired a cat through a set of circumstances I don't need to get into now, and I started to realize that I didn't really hate cats. In fact, I realized quite soon that I didn't really know what cats were all about. I didn't have any appreciation for them. I had spent two years living with my roommate's cats and "tolerating" them, but I had been hesitant to warm up to them. But with my own cat, the more I got to know Pickles, the more I started to really appreciate her. Cats are independent and incredibly inquisitive, and they are more agile with more dexterity than dogs, all of which makes them fascinating creatures.

I was forced to admit that not only didn't I hate cats, but I might actually like cats. I was also forced to admit that my long-standing conviction that I hated cats was not mine; it was my father's. Just as we inherit genes, behaviours, tendencies, and money, so too do we inherit beliefs, and I had inherited this one — along with many others — from one of my parents. He too came by it honestly, for his own father hated cats. Although, as I adapted, so did my dad. One summer I went home to Winnipeg to stay with my dad for a month, and I brought both Jett (my dog) and Pickles with me. It took some convincing that this arrangement was a good idea, but we had a great time together, and my dad even began to appreciate cats — or at least Pickles — too. And my grandfather? Well... if we'd had more time, he might have come around, but he definitely became more curious about Pickles, which in hindsight is not that surprising to me because he too is an animal lover.

Seeing how ignorantly I had adopted a belief that went against a deep love for animals forced me to look more closely at my other beliefs. As I consider each belief, I am forced to consider its source, be it my father, my mother, my friends, my mentors, my culture, or simply a combination of many elements accumulated in my almost four decades of having a malleable mind. In the process I also realized that I didn't hate Winnipeg, the city I grew up in, but I had noted that most other people say they hate Winnipeg, so I just adopted that belief too.

I don't hate cats and I don't hate Winnipeg, but I was led to believe that I did. This happens all the time. Our thoughts are influenced by many circumstances around us and we should put a great deal of care into considering our thinking patterns and how they influence us because our thoughts are deeply personal. Our thoughts resonate in our bodies, minds, brains, and behaviours. Consider the effects of negative versus positive thinking.

Our Thoughts Can Hurt Us

While sitting in an airport lounge one day, I was drawn in to the many disturbing stories populating the news. Gloom was everywhere I looked, and I couldn't help but notice how my body reacted to all of this. I felt nauseous, filled with rage. Blood rushed through my veins, my heart was beating fast, my eyes were scrunching, tension grew in my head, and my shoulders lifted up in defence. Then feelings of helplessness emerged. My thoughts, my perceptions, manifested in my body. My sympathetic nervous system, my amygdala, and my insular cortex were all active, causing negative experiences inside of me. My body began to feel limp, heavy, defeated, and almost paralyzed. I felt helpless to do anything about the suffering of people I was reading about. These stories don't just linger in our minds; they are in our bodies and our brains. They get incorporated. They are embodied.

From a neuroscience perspective, our brain can easily be primed. This has been discovered through the use of electroencephalogram (EEG), when electrodes are placed on a person's head to visualize the electrical activity of the brain or brainwaves. When brainwaves are being recorded, researchers can present information to people (and their brains) and evoke different kinds of brainwaves that are indicative of different states of awareness and information process. These are known as "evoked potentials" or "event-related potentials" (ERPs). These ERPs have shown that the brain can be primed to respond in a certain way. For example, in one experiment researchers asked people to respond by pressing a button to indicate left or right after they were shown an arrow that pointed in one direction or other. But just before the arrow was shown, the participants were presented with a very short (almost subliminal) cue that either indicated the direction about to be shown or the opposite direction. What the researchers observed in this study was that when the cue beforehand predicted the direction the people were going to respond to, their brains responded quickly. In fact, people (and their brains) responded more quickly than if there was no predicting cue. This showed that small cues in our environment can affect how the brain responds in the future.

The body reacts with a variety of changes connected to the fight-or-flight response, or stress response. Changes in breathing, heart rate, skin conductance, and blood pressure have all been documented in relation to the presentation of emotionally negative words. Try it for yourself now.

Exercise

Consider how you feel in your mind and body when you review these news story headlines:

Boy Dies in Fire

Crisis Continues Overseas

Millions Set to Lose Their Homes as Economy Continues to Crash

Seniors Lose Millions in Investments and Are Forced Out of Retirement

Fear Continues to Loom after Recent Terrorist Attacks

Notes:

We know that our brains do indeed respond differently to different types of stimuli. For example, in one study,[21] positive and negative words activated the amygdala (the structure known for its role in emotions). But positive words also activated the ventral medial prefrontal cortex whereas negative words activated the insula (or insular cortex), an area involved in bodily self-awareness, emotion, and disgust.

Similarly, our brains respond in a particular manner when presented with death-related words, which often have a negative connotation in our culture. When participants of a study were shown death-related words, they had increased neural activity in areas of the frontal-parietal cortex[22] and decreased activity in the insular cortex.

Words associated with pain also activate specific areas of the brain.[23] When participants in another experiment were asked to imagine the experience associated with words (either painful or neutral) presented on a screen, only pain-related words activated areas of the frontal brain.[24] In the second part of this study, the researchers presented the participants with a distracting task while words associated with pain were presented on the screen

21. Thomas Straube, Andreas Sauer, and Wolfgang H. R. Miltner, "Brain Activation During Direct and Indirect Processing of Positive and Negative Words," *Behavioral Brain Research* 222 (2011): 66–72.

22. Specifically, the inferior parietal lobule, the right frontal eye field, and the right superior parietal lobule: Zhenhao Shi and Shihui Han, "Transient and Sustained Neural Responses to Death-Related Linguistic Cues," *Social Cognitive and Affective Neuroscience* 8, no. 5, (2013): 573–8.

23. M. Richter, W. Miltner, and T. Weiss, "Pain Words Activate Pain-Processing Neural Structures," [article in German], *Schmerz* 25, no. 3 (2011): 322–324.

24. The dorsolateral prefrontal cortex (DLPFC) and inferior parietal cortex (IPC).

again to see if they could ignore the words. Nope. The words associated with pain continued to affect areas of the brain, particularly in the cingulate cortex.[25]

We also see a variety of heightened or dampened responses to emotional words in people with depression, schizophrenia, psychopathy, and those with eating disorders. In another study[26] of individuals diagnosed with post-traumatic stress disorder (PTSD), emotional words created greater activation in certain brain areas.[27] In fact, these people tended to overreact to the presentation of negative words in a manner that corresponded with heightened activity in the brain.[28] Knowing what we do about PTSD, it seems quite possible that individuals who suffer from it are experiencing a primed or exaggerated response to emotional and negative triggers, one that is reflected in their behavioural, emotional, and neural reactions.

In another interesting study about how the medical system delivers information to patients,[29] an MRI report was reworded. The purpose was to see if results could be delivered in a manner that reduced stress and emotional responses in patients[30] and certain words, such as "tear"

25. The dorsal anterior cingulum was decreased and the subgenus anterior cingulum was activated.

26. Kathleen Thomaes et al., "Increased Anterior Cingulate Cortex and Hippocampus Activation in Complex PTSD During Encoding of Negative Words," Social Cognition and Affective Neuroscience 8, no. 2 (2013): 190–200.

27. Left ventral anterior cingulate cortex (ACC) and dorsal ACC extending to the DLPFC.

28. Mistakes were positively correlated with activity in the left ventrolateral prefrontal and orbitofrontal cortex.

29. Jeroen K. J. Bossen et al., "Does Rewording MRI Reports Improve Patient Understanding and Emotional Response to a Clinical Report?" Clinical Orthopaedics and Related Research 471, no. 11 (2013): 3637–44.

30. The aim was also to improve understanding on behalf of the patients by using more accessible language.

were eliminated. The result of reframing the language to be more positive was encouraging and provided a much better experience for patients reading their own reports.

All of these studies (and this is by no means an exhaustive list) go to show that words are not benign. They affect us. They affect the brain. They also affect the body. They affect the mind. They can even affect emotional well-being, which can affect health.

Exercise

Compare how you felt when you read the previous headlines to your feelings when reading the following:

Dog Saves Drowning Boy and Is Honoured with a Big Bone

Community Comes Together for Multi-Faith Holiday Festivities

Anonymous Pay-It-Forward Acts of Kindness Are Sweeping the Nation

Billionaire Donates 99% of Wealth to Charities

Now, ask yourself if anything feels different.

Notes:

Brain Filtering

Although thinking positively or negatively and switching from one to the other is possible, adopting a particular mindset (or making this choice) is not necessarily easy. We spend a significant amount of time teaching the brain what to focus on, and that focus becomes our reality. We essentially set up neural filters that influence the weight we give to different pieces of information. If we spend lots of time thinking negatively, we will be primed toward the negative, and the brain will exclude the positive or see it as less important content. If we spend lots of time thinking positively, then we will not only be primed to see the positive, but our filtering system will favour positive over negative. The brain functions on a premise of trying to make things efficient, so it tends to perpetuate a way of thinking, following the path of least resistance. Whether we want that path to be positive or negative is somewhat within our control.

A perfect example of how our filters affect what we see comes from a woman in one of my life-coaching courses. She was experiencing an episode of depression to the point where she was quite negative with herself for the first few months of the course. At the outset of our fourth course weekend together, she was on the upswing and obviously much more positive. She even knew it herself. We had been meeting as a group in the same room for all of the sessions, and right outside of the classroom was a chalkboard where people had written comments. In the middle of the board, someone had inscribed *Be Happy* in large letters. I had noticed it many times during the course weekends, but on this particular day the student who had been experiencing depression walked out of the classroom and said enthusiastically, "Oh look. Someone wrote *Be Happy*!" I laughed at first and then realized she had not actually seen it before. When I told her it had been there since the beginning of the course, she was astonished. She looked at the board in disbelief and pointed to other words (that were not particularly positive or negative) and said, "No. These words were there but the 'Be Happy' was not. I swear!" I eventually convinced her that, indeed, they had been there the whole time and that perhaps her sudden noticing of it was reflective of her mood lifting and filters changing. Having already covered this concept in the course, she let herself be convinced. But seeing someone so obviously ignore, or filter out, positivity was incredible to me.

Filtering Positive Versus Negative

While attending the 2014 Canadian Ultimate (Frisbee) Championships in Waterloo, Ontario, we were met with awful weather, at least according to me. It was August in Ontario, so the expectation was for it to be warm (at least 25 degrees Celsius), but instead it was unseasonably cold (10–15 degrees), really windy (which is not great for Frisbee-throwing), and rainy. I happen to really enjoy playing in hot weather — like, 30-degrees hot. Learning that this tournament was going to be awful weather would normally send me into a three-day frenzy of complaints. Just before we hit the field for our first game, I caught myself spiralling down. I turned to my teammate and said, "I'm not going there. From this point on, I'm not allowed to complain. Not until Sunday evening when it's all over." Indeed, with that statement I committed to my goal of thinking positively, or rather refraining from negative thinking, which at the time was more realistic.

The results were actually amazing. I had a wonderful time. Granted, the weather was horrible and there was so much I *could* have complained about, but I knew that way of thinking would not serve me and, if anything, would more likely hurt me.

Now, if I had been able to do something about my negative situation, then indeed I should have done it. When we find ourselves in negative situations we can change, we should seek the courage to do so. Get out, leave, speak up, make a change, or do whatever we can to get out of the negative situation. But, if we cannot, should not, or will not take action against our negative thinking, then we might as well find something positive in it, or accept it and move on. The serenity prayer says it quite nicely: *God grant me the serenity to accept the things I cannot change; the courage to change the things I can; and the wisdom to know the difference.*

The filtering system is not a bad thing. It helps us deal with the vast amount of information the brain is exposed to moment by moment. The brain processes much more information than we are consciously aware of. As a result, much of it doesn't pass through the filters as being relevant enough or salient enough to bid for our attention. Imagine, for example, that your brain has the task of surveying your external and internal environments and alerting you to that which it deems important. The brain is forced to make quick decisions on what to let go of and what to pass on to upper management. How does it make those decisions? It's a recruitment of resources issue. If I'm looking for all things bad, then my brain will deem those important and pass along that information while filtering out the good. I've told my brain that good things are non-essential and that we don't need to devote any resources there. Remember, the path of least resistance is more efficient and the brain prefers to work efficiently. If, instead, I tell my brain that all things good are my focus, then my brain will deem those important and pass that information along to me. It's really no different than the filtering system we sometimes use to quickly scan material (for example, a pile of resumés). We prime our brain with what is relevant and then we scan quickly to connect what is relevant with the information we are presented. That's essentially what we are doing with negativity versus positivity.

Although thinking positively or negatively and switching from one to the other is possible, it requires practice. In practice, we often spend our efforts in positive self-talk. We utter the words "I should see the positive." Or "I shouldn't dwell." Or even, like the *Saturday Night Live* character Stuart Smalley used to say, "I am good enough and smart enough, and gosh darn it people like me!" Many of us know that if we focus on negative things, we will bring in negative things. This is the law of attraction. Dwelling and wallowing in negative states is not healthy. Being positive is better for our health, disease outcome, and the prevention of relapses of depressive episodes.

When we are in a state of negativity, we often literally cannot see the positive things around us. Our brains become primed toward the negative, filtering out the positive and letting in only negative. Priming is a well-studied phenomenon in psychology. Here is a simple example of how you can be primed to say a particular word. Read these two sentences aloud and then answer the question:

At Hallowe'en there are lots of witches and ghosts.

I'm about to have a dinner party. I'm excited to be the host.

In the morning, what do you put in the toaster oven?

———

You probably said "toast," right? If so, I primed you to say that with the other words that sound like toast. But upon further reflection, is "toast" really what you put in the toaster, or is it bread?

I could ask you to read over the list of words below once:

grocery	produce	cart	cashier
magazines	eggs	milk	bags
meat	cereal	pop	potato chips
line-up	mart	bacon	coupon
cheese	aisle	roasted chicken	

Now, cover that list of words so you cannot see them. Of the words listed below, which were present in the original list?

coupon
cashier
shopping
cart
all of the above

Did you pick "all of the above"? If so, you did what the majority of people do wrong. They make a quick judgment about a word being present because it is thematically associated with the first list. It's a quick way for the brain to deal with lots of information and to avoid exerting a lot of energy for memorization. You can take a best guess and probably be right, except in a situation designed to trick you, like this one!

Another good example of filtering can be seen in some classic psychology experiments conducted by Daniel Simons and colleagues. Have fun with some of videos posted on his website, www.dansimons.com/videos.html.

Gratitude as a Filter

Gratitude journals are all the rage among the personal development folks, the positive thinkers, and life coaches in particular, and for good reason. In essence they are an easy self-directed neuroplasticity positive-thinking filtering strategy. Writing in a gratitude journal is a simple strategy to evoke in your brain what you think is important in your mind, brain, and body. Researchers have the luxury of using sophisticated tools to examine when the brain is being evoked (like ERPs). Outside the lab we have to trust that our brains are being evoked even when we can't see it directly. But we can evoke our brains.

The typical task associated with gratitude journals is to write down three things we are grateful for each night before going to bed or each morning upon rising. For some, this might be tough; they might come up with only "comfy pillow," like one client I had who was suffering an episode of depression. For others, it is easier. And some days may be easier than others.

Gratitude Exercise

In one minute (timed), write down as many things as you can that you are grateful for.

Do this twice a day for five days, making note of how you feel each day.

Compare how you feel at the beginning and end of these five days.

Notes:

The practice helps us redirect and prime our brains toward the positive. In studies, it has been shown to be helpful. People who adopt a practice of gratitude journaling report a greater sense of wellness and life satisfaction. I took the practice a step further by starting to really celebrate Thanksgiving, sending out gratitude cards in place of any other seasonal cards (like Christmas cards). I felt that the gift of gratitude was not given enough, and I was happy to offer it in celebration of an already-in-existence holiday!

Optimism — A Related Learned-Thinking Skill

We weren't born thinking negatively or positively, at least as far as we can tell. We were born with very adaptable brains that respond to the environments in which we are placed. As we grow, we adapt to our environments and learn what adaptive thinking is and what it is not. Dr. Martin Seligman wrote an excellent book called *Learned Optimism*,[31] where he describes the history of his own thinking as a psychologist who went from seeing the negative to seeing the positive, in his own life and as a lens for the entire discipline of psychology. As a proponent and the founder of positive psychology, he noticed that much of the work within psychology is devoted to the study of the abnormal. Most courses and studies are focused on the ways human and other animal minds have gone wrong. Little is devoted to how our minds thrive. As a result, we have been ignoring the vast potential to understand how humans thrive, at least until Seligman put forward his positive psychology manifesto in 1998. It started with an understanding of how laboratory animals learned to become helpless, and then flipping that around.

Seligman was studying learned helplessness, one of several defining characteristics of depression, and still studied a lot today. When I was doing my doctoral research, we used several paradigms to invoke a sense of helplessness in our laboratory animals as a model of depression.

In the learned-helplessness model, laboratory animals are placed in a situation from which they cannot escape.[32] This might be a water tank with no way out; it may be in a room with a continuous loud noise; or it could be physical restraint where struggling does nothing to promote freedom. Regardless, in each situation the animal typically

---

31. Check to see how optimistic you are at
http://www.stanford.edu/class/msande271/onlinetools/LearnedOpt.html

32. J. Bruce Overmier and Martin E. P. Seligman, "Effects of Inescapable Shock Upon Subsequent Escape and Avoidance Responding," *Journal of Comparative and Physiological Psychology* 63, no. 1 (1967): 28–33.

learns that it cannot escape despite its best effort, and the animal must endure the negativity of the situation. Learned helplessness is not good for the mind or body. It has been associated with faster tumour growth in laboratory animals and a greater likelihood of transplant rejection. On the contrary, learning to escape has also been shown to "immunize" young animals against cancer, for example.

The initial negative situation (not being able to escape) is not, itself, the problematic part. The animals are given no means of escape, and this is analogous to accepting the things we cannot change. What is interesting is the way an animal that previously learned to be helpless responds in future situations. After some amount of time, the animals are returned to an environment very similar to their previous situation, when they learned there was no escape. However, this time, there *is* an escape option. The animals could get out of the water tank, turn off the loud noise, or exit their restraints. Unfortunately, animals that previously learned there was no escape do not even bother attempting to escape or looking for an escape option! Instead, they sit there and whimper, roll over, or simply do nothing at all. In some examples animals are in small boxes where they could easily jump over the walls to escape but they do not. They have learned to see no option. Strong filters have been set. This is where the animals failed in the wisdom to know the difference between what they could and could not change.

In another study,[33] human participants were asked to sit and wait in a room. Loud, obnoxious sounds blared through speakers at both groups. For one group, there was a way to turn off the sound. For the other group, there was no way to turn off the sound, and they were forced to sit and endure the negative situation. A few days later, both groups were returned to another room with a loud sound and this time both groups had the opportunity to escape the annoyance. The first group did not hesitate to find the escape mechanism (a switch on the wall). But the second group sat there, knew the switch was there, but did nothing to escape the situation. They had essentially learned to be helpless, just like the laboratory rats and dogs.

The existence of learned helplessness is absolutely incredible, and it should provoke us all to think about our own circumstances. When do we become helpless and when do we not become helpless? Obviously this phenomenon does occur in humans.

Learned helplessness is an example of a natural brain-mind-thinking tendency gone awry in the sense that our brain seeks the easy, least-resistance path. We don't often want to exert unnecessary energy, especially once we've learned that our efforts do not amount to anything. The problem, of course, is when we fall into passive acceptance and fail to see the opportunities to escape, or more broadly, we fail to grow, rise, and thrive in opportunity.

I see such failures happening a lot, particularly when people find themselves in jobs where they feel like they have little ability to take control and escape something

33. Donald S. Hiroto and Martin E. P. Seligman "Generality of Learned Helplessness in Man," *Journal of Personality and Social Psychology* 31, no. 2 (1975): 311–27.

negative. The same failures occur for people in jobs they feel they can't leave; people in highly bureaucratic environments (education, government); and people living in abusive relationships. I get a sense of it teaching at the university because of the incredible demand to please the students, the need to demonstrate great teaching evaluations, a need to keep grades from creeping up, the need to deliver excellent and engaging lectures and learning opportunities with few resources to do so, and the need to satisfy other administrative requirements. But we can change our brains and train ourselves into thinking opportunistically, positively, and optimistically, and that makes a world of difference.

Shifting his own mind as a psychologist, Seligman began to see the potential in psychology to study the ways humans could thrive. Within the area of learned helplessness specifically, Seligman realized that if people and laboratory animals could be taught to experience learned helplessness, perhaps they could also be empowered to see opportunity. This opened a door to the concept of "learned optimism," which became the title of one of Seligman's books, and much more broadly, the concept of positive psychology, the study of human and community thriving.

In his book, Seligman talks about how using cognitive therapy (mental shifting or reframing) is effective for switching people's pessimistic views into ones that were optimistic. This is particularly important for people with depression. Depression is often accompanied by many ruminating thoughts of worthlessness, hopelessness, purposelessness, and helplessness. Antidepressant medication can, at times, effectively manage some of the negative experiences associated with depression, but it does little to change a person's ways of thinking. The problem with treating depression only pharmaceutically is that when people come off the medication, they can easily relapse. However, those who undergo cognitive therapy and learn to change the structure of their thoughts and their brain's filtering system show improvement from depression. When trained to think differently, these people are also less likely to relapse into future episodes of depression.   Many of us have experienced depression. Many of us have been medicated or otherwise sought treatment for depression. Many of us will experience depression in the future, but learning to think optimistically and positively has benefits far beyond relief of depression. It really is about learning to thrive in mind, body, and brain. There is good scientific evidence that those who are optimistic are more likely to succeed (in school, in work, and in elections), have better health prognoses, and live longer.

Changing Our Minds, Changing Our Brains

Our filters are not permanent, thanks to neuroplasticity. We can change the path of least resistance if we choose, not overnight mind you, but we can change our ways of thinking. In order to actively change the mind and brain, we must first observe what it's doing now so we have a starting point.

Observing Our State of Mind

The observation of our state of mind and the contents of our mind is an important part of clearing away negativity and bringing in positivity. Am I mad? Am I sad? Am I irritated? Am I frustrated? What am I? The goal first is to just observe our current state without trying to alter it.

Mind Observation Exercise

Try to pay attention to your breath and only your breath for a full five minutes. Set a timer.

Make mental note of any thoughts that come up while trying to focus on your breath.

Afterwards, check to see if any of your thoughts looked like any of these:

I feel like I'm up against a wall.

I'm not good enough.

Why can't I ever succeed?

No one understands me.

I've let people down.

I don't think I can go on.

I wish I were a better person.

My life's not going the way I want it to.

I'm so disappointed in myself.

I can't get started.

I can't get things together.

I wish I were somewhere else.

I hate myself.

I'm worthless.

I wish I could just disappear.

I'll never make it.

These are thoughts reported by people with depression as listed in the book *Mindful Way Through Depression: Freeing Yourself from Chronic Unhappiness*.[34] Here are some that plague my brain:

I'm so stupid.

I don't have enough money.

I don't have enough time.

I wish I had more time for yoga.

I don't feel like going for a run.

I am lazy.

I am fat.

Why do I bother?

I wish someone would take a chance on my crazy ideas.

Notes:

---

34. J. Mark G. Williams, John D. Teasdale, Zindel V. Segal, and Jon Kabat-Zinn. *Mindful Way Through Depression: Freeing Yourself from Chronic Unhappiness*. (New York: Guilford Publications, 2007).

They may not all appear negative to you but I can assure you they are. When I say them to myself, I don't feel good. They make me feel bad, anxious, defeated, and they cause me to experience all of the body sensations described at the beginning of the chapter when I was reading the newspaper.

Take another two minutes to generate more negative thoughts that you identify with, thoughts that you personally say to yourself that are not positive or neutral. They don't have to come from the list you just generated. It can be something that you are only thinking of now. Or feel free to borrow some of the ones you have seen. They are by no means any one person's intellectual property.

Sometimes we need to dig deeper to uncover the negativity behind a thought. Take my thought, *I don't have enough time.* On the surface it does not look negative. But if we simply look at the syntax, it *is* negative. The word "don't" implies negativity. This is the rationale for a common parenting practice where parents refrain from using sentences that say "no" or "don't." For example, if a child goes to grab something that the parent feels is inappropriate for whatever reason, instead of saying "No! Don't touch that," the parent might say "leave that" or "here, let's play with this instead." I have a dog and the same logic applies. I tried long and hard to tell my dog not to bark and little changed. Now instead of saying "No! Don't bark," I grab my dog's attention with a trick or another command she knows, and that she knows she will get a treat for. I also get her to drop things out of her mouth by essentially saying, "leave that and try this instead." It's just a simple rewording, but that's often what being positive comes down to, especially when considering how we speak to ourselves.

Have another look at your list and see if anything there has either "no" or "don't," either explicitly stated or just implied. Can you go ahead and reword without using any negative words?

It's Personal

What else do you see in your list and the lists provided above? Can you identify any other patterns? Did you notice that there are several thoughts that include the word "I"? Indeed there are. We make these thoughts really personal, and we believe them because they feel personally generated. I love how authors Williams and colleagues put it: "Negative thoughts are part of the landscape of depression. They are symptoms of depression just the same as aches and pains are symptoms of the flu. There is nothing personal about them."

That statement is meant to be applied to those with depression, but I read it much more generally, regardless of whether we have been diagnosed with depression or not. Indeed, negative thoughts are part of the landscape of being *human*. They are symptoms of the mind just the same as aches and pains are symptoms of the flu. There is nothing personal about them. But we take them personally, at least until we observe ourselves from a distance like we did in the last exercise.

How We Explain Things to Ourselves

Overly identifying with our thoughts can be a problem even if we are not depressed. Several scientific studies have shown us that the way we explain things that happen to us predicts our likelihood of developing depression, our health prognosis, and the success we experience. This is referred to as our "explanatory style," and it predicts whether we are optimistic or pessimistic, as Seligman describes in *Learned Optimism*. When examining our explanatory style, we should look for the 3 Ps: Personal, Permanent, and Pervasive.

Personal refers to the "I" already described above. When bad things happen, people who are depressed or pessimistic tend to attribute them to something they have done themselves, or to some characteristic of themselves, partially representing their low confidence. Alternatively, when good things happen, they attribute them to some external cause rather than to themselves, resisting the opportunity to claim achievements. In many ways this is illogical and rather inconsistent. How can bad things be our fault but good things not be? It really isn't reasonable, but our thinking is not always reasonable. Comparably, optimistic and happy people attribute bad things to circumstances rather than to themselves, effectively reducing the impact on their ego. Yet they are also able to claim good things, contributing to positive self-image. That may not be reasonable either, but it does correspond with greater happiness. Of course a balance is necessary so we don't inflate our egos and fail to take responsibility for anything we do that goes wrong.

Permanent refers to the inherent belief that something that is there will always be there. For example, when looking at an obstacle, a set-back, or a reason for something bad that happened in their lives, people with depression and pessimistic thinkers attribute the bad thing to some permanent feature in themselves. This is evidenced by the phrases like, "I am worthless" or "I am lazy." "I" is personal, and "am" is permanent. A statement like this implies now and forever. There is no evidence of it being a past experience or something that could or will change for the future. Imagine how defeatist that feels. Permanence implies no possibility for change, no way out, no opportunity. On the other hand, seeing opportunities by acknowledging impermanence is a feature of optimism. Optimistic people see that things are always changing. For example, an optimist may say "that was a lazy day," or "that was a lazy way of doing that," implying that the laziness does not have to be true tomorrow. As a result, by acknowledging that everything changes, optimistic people have no reason to feel stuck in a situation.

Pervasive refers to the global, all encapsulating state of affairs. For example, with the phrase "I am a failure." There is a big period at the end. The period stops the thought short of any other consideration that the experience could be circumstantial. Instead, I believe I am a failure, permanently and in all aspects of my life.

Optimists think differently. They see a failure as being attached to a circumstance, not a global characteristic of themselves. A pessimist might believe they are a failure because they didn't get an interview; whereas an optimist would equate the situation with, for example, not having time to individualize a cover letter, or not fitting in with what the company was looking for.

This explanatory style is a basis for how to change our thoughts into something positive, optimistic, and empowering. The goal is to allow our thoughts to become agents for change and action rather than obstacles that create stagnation, defeat, and helplessness.

Changing Our Thinking

Here is another common example of a thought that needs transformation: "I can't." Again, it's personal ("I"), and it's permanent because it implies forever. The statement might also harbour a bit of pervasiveness. Maybe we are referring to asking for a raise or going back to school or writing a book or being a good parent. We might feel like we cannot do something, but more likely we cannot do something under certain specific circumstances. Perhaps we cannot ask for a raise until the recession passes or until we have been employed for two years. Similarly, we may not be able to go back to university, but perhaps we could take a course offered online or through a community centre. "I can't go back to school" is a broad statement, but "I can take this really interesting course about neuroscience and life skills and yoga" is much less daunting. It is opportunistic and specific enough to be achievable. Similarly, we may feel unable to write a book from cover to cover, but we may be able to write small chapters or paragraphs that might eventually turn into a book, or we might start blogging and see where that leads us. Essentially, what we aim to do in shifting our perspective is to focus on what we *can* do rather than on what we cannot!

The other big thing to watch out for in observing, analyzing, and reframing our thoughts is our tendency to assume we can predict the future. Despite my sincere interest in super powers, I have yet to see evidence that our brains can predict the future. But we imply that we can predict the future with statements that assume "always" or "never," and those (probably) false predictions resonate in our bodies and linger in our minds just like other negativity does. We believe in those thoughts because we take them personally. Limiting what words make up our thoughts can transform our thinking. When we remove negative language, we take unhelpful, negative, and ruminating thoughts out of our mind and create actionable opportunities for change. But this takes work. We can't just transform a new thought on a piece of paper and then walk out the door and expect life to change. We actually have to work on it. We need to change the brain by changing the filters and priming it with intentional positivity. It requires careful attention and frequent check-ins with our thinking patterns and possibly with our life coaches. If we do so, we teach our brains to be optimistic, and with optimism comes health, happiness, long life, and success.

The Neuroscience of Optimism

Humans are prone to be optimistically biased when it comes to predicting future events. For example, we typically underestimate the likelihood of negative events occurring in the future. This bias seems to be a result of a frontal-lobe system that is wired to update itself when tracking positive events (resulting in a more realistic estimation for the future), but does not update occurrences of negative events (resulting in an underestimation for the future). The result is that the brain (frontal-lobe system) generally becomes good at predicting the likelihood of future positive events but bad at predicting the likelihood of future negative events.

This "optimism bias," as it has become known, is associated with levels of a particular neurochemical, dopamine, which is also involved in experiences like motivation and reward. In a well-designed study published in 2012 in the scientific journal, *Current Biology*, researchers Sharot and colleagues[35] tested to see what would happen to people's predictions of negative events if those individuals' dopamine levels were manipulated. The researchers made the adjustments by administering L-Dopa, a synthetic precursor of dopamine. Administering L-Dopa is known to raise dopamine levels. Once dopamine levels had been manipulated, the researchers asked participants to predict the likelihood of 40 different negative events,[36] a list that included examples like fraud when buying something on the Internet, theft from vehicle, sport-related accident, rodent in house, knee osteoarthritis, dementia, drug abuse, being convicted of crime, having fleas/lice, sexual dysfunction, severe teeth problems when old, and cancer. Next the researchers provided the participants with the actual probabilities of the occurrence of these events, and then they readministered the list, asking again what the participants felt the likelihood was of personally experiencing any of those negative events. The results of this study were clear: higher dopamine levels biased people toward more optimistic predictions at the beginning, and people did not adjust their underestimation of negative events even after confronted with real probabilities. Essentially, when stimulated with extra dopamine, they maintained their

35. Tali Sharot et al., "How Dopamine Enhances an Optimism Bias in Humans," *Current Biology* 22, no.16 (2012): 1477–1481.

36. For a full list of the negative events, go to http://www.cell.com/cms/attachment/*2021739663*/2041550901/mmc1.pdf

optimism that bad things would not happen to them, despite seeing the odds.

Manipulating dopamine levels does not just happen in scientific studies. In fact, we are often manipulating our neurotransmitter levels by virtue of what we eat. Another way to boost dopamine is through consuming tyrosine-high foods and drinks. Tyrosine (an amino acid) is a natural precursor that is found in protein-rich foods. It is eventually converted into dopamine so the body can use it as the transmitter and hormone that it is. Studies show that boosting dopamine by increasing tyrosine-rich foods does affect other aspects of our thinking, but I am aware of no study to date that has dealt specifically with whether these foods boost optimism. Still, there is good reason to believe they would.

When Not to Be Positive or Optimistic

Optimism does come with benefits to our mind, body, and brain. But should we always be optimistic? No. There are certain circumstances in which we do not want to be overly optimistic. For example, suppressing emotions creates more negativity,[37] which is associated with depression, anxiety, obsessive–compulsive disorder, and overeating, exacerbates disease progression, and can alienate people into feeling "fake." Being aware of our thinking patterns gives us power to decide when to invoke the power of positivity, optimism, and thought transformation, when to experience our real emotions,

---

37. For a full and updated list, head to the U.S. National Library of Medicine's PubMed and search "suppress and negative emotions" (http://www.ncbi.nlm.nih.gov/pubmed/?term=suppress+and+negative+emotions).

when to accept what cannot change, and how to know the difference.

Consider someone, maybe you, who has ever felt the kind of intense misery that weakens us at the knees and seems to crush our chests. Perhaps we just learned the devastating news of a serious illness, lost a loved one, or were fired. Perhaps we had a really bad argument with a parent, or our significant other just broke up with us. Perhaps we just realized how badly we hurt someone else and we feel awful for it. In these moments, we are anything but positive. Living may seem futile and, in that moment, life may feel hopeless and worthless. It is incredibly hard to hear the words "be positive" or "cheer up." Why? Because as powerful as the mind might be, without adequate training it may not be able to overcome something as severe as despair or depression. The body is undergoing an immediate physiological faction. Mind over matter is not appropriate here. In that moment we are feeling the sensations within our body. It's only after the moment has passed and if we continue to ruminate that we would want to invoke the power of optimism. If we ignore the physical and psychological symptoms that we experience in that moment, then we are creating a mind-body disconnection. We are essentially teaching the mind to ignore what the body is saying, and that practice can perpetuate a filtering system where the brain learns that what is going on in the body is not important, not relevant, not worthy. We learn not to pay attention to messages from our body. This becomes a problem when we need to pay attention to the messages of pain or illness, but the brain has adopted filters that exclude this information.

The other risk we run in "being positive" or "cheering up" is the loss of information about our situation or state. It becomes an all-or-none phenomenon with little insight into the in-between. If the body gets a virus, do we say "I don't have a fever"? If we break a leg, does anyone say "don't be sore"? If we cut ourselves, do we say "don't bleed"? No. These are important messages being delivered by our body. Invoking positivity and optimism is less important than realism. Yes, there may be times when we need to dissociate from pain in order to survive, for example, if you were lost in the woods with a broken leg and needed to make it to safety, you might benefit from mind over matter. But otherwise, a sore broken leg keeps us from using that leg while it heals. Similarly, a fever, although unpleasant, is working for us. A bleeding cut, although unpleasant, tells us that we need a physical repair. In the same way, if the mind feels sad, why would we say "don't feel sad"? If the body gets cancer, why would we say "cheer up"?

It's worth mentioning that there are benefits to not being overly optimistic. For example, pessimistic people, although typically more depressed and less healthy, are actually wiser and make better judgments, particularly in times of high risk. Gauging whether we can jump from one cliff to another is not a time to be optimistic. Neither is it advisable to be optimistic in deciding to walk up to a growling dog. It is similarly not advisable to be optimistic at the casino if we are about to gamble away our last $100 (or our house). Being realistic and even pessimistic is adaptive.

I heard a great example of when not to be optimistic during an interview with Canadian astronaut Chris

Hadfield. He spoke about his training, which included in-depth thinking about what could go wrong during his space mission. His rationale was that he knew things would go wrong, and he needed to prepare his brain to act, not think, in the moment so that he could survive. The smallest of errors could cause him to die, so he needed to be aware of all that could possibly go wrong. My dad was a linesman for Manitoba Hydro, where similar precautions were necessary. Working around high voltage circuits is not a place for optimism. Believing that he would probably be okay if he cut the wrong wire would not have served him. Instead, my dad, like Chris Hadfield, would think about all the ways things could go wrong. The downside to this, in my opinion, is that it is hard to turn off that way of thinking outside an environment where pessimism is adaptive. I remember my dad telling me many times about what I did wrong rather than what I did right — in fact one of those times was right after he attended one of my workshops on positive thinking!

In summary, optimism is great for the following situations:

- when we are in an achievement situation or anytime we need to be successful (getting a promotion, selling a product, writing a difficult report, winning a game);
- when we are concerned about how we will feel in the future (fighting off depression, keeping up morale);
- when the situation is apt to be protracted and physical health is an issue; and
- when we want to lead, inspire others, and win votes.

Optimism is not great in these situations:

- when our goal is to plan for a risky and uncertain future;
- when our goal is to counsel others whose future is dim; and
- when we want to appear sympathetic to the troubles of others.

Despite these few caveats, the practice of positive thinking and optimism is generally a beneficial one. I remember having several conversations with a client about this as I was attempting to convince her of the benefits. She hung on to the idea that optimism was not seeing things as they really are. In fact, that is probably true. Seeing things as positive and wonderful amid the world full of torture, war, poverty, and so on is probably ignoring some important facts. But seeing the world as a black, gloomy, doom-filled place is equally unrealistic when we know there are many amazing people doing good work that is incredibly positive. For the most part we don't see the world, ourselves, or our lives as they truly are anyway. Our minds are already misconstruing several facts of our existence. But therein lies the choice of how we want to see the world, recognizing it is within our control.

If we observe (and choose) our state of mind, then we are able to build new and desired filters. We are better able to see that which is truly negative and that which is truly positive. By knowing the truth about our negative thoughts, we are able to refine our filters. A filter that is porous at first becomes more selective as it learns what is negative and what is positive. And then with almost no effort at all, negativity loses its strength and begins to fall away. It no longer dominates us or enters our spirit unknowingly. Negative is caught at the filters. It is at this point that we can understand the benefits of being positive. We are truly living in a state of positivity and are no longer sensitive to the effects of negative thoughts, just as I attempted when I chose to not complain about the weather at my ultimate tournament. If we spend too much energy trying to alter ourselves from our natural state into something positive, we may miss out on important information being delivered. Although unpleasant, finding ourselves in a negative state may tell us important things, like we're in the wrong job, in the wrong relationship, going after the wrong dream. A negative state lets us know we may need to work on some piece of our life. In that light, seeing experiences as they are is never a bad thing. It's only bad when our way of thinking keeps us from seeing solutions out of our particular mess.

Becoming a Positive Thinker

Giving and Receiving Positive Feedback

Having done several degrees in psychology, I understood the concept of positive reinforcement as a principle of behaviourism. The idea is simple: reinforce behaviours you want to see again. Positive reinforcement is literally about increasing the likelihood of a behaviour by providing a rewarding stimulus. So for example, if I want my dog Jett to sit in the future (behaviour), I will reward her sitting behaviour now with a treat (reinforcement). That should reinforce sitting and increase her likelihood to sit in the future. Comparatively, negative reinforcement is also strengthening the likelihood of a behaviour, but this time, it is achieved by removing a negative or unwanted circumstance. For example, if doing yoga (behaviour) reduces tension in my body (unwanted circumstance), then the likelihood of doing yoga again is increased by the removal of the unwanted tension. Negative reinforcement is often confused with punishment, which is different in that punishment is an attempt to decrease the likelihood of a behaviour. We are all familiar with punishment. It happens when we are penalized for being late, missing a deadline, or caught speeding, for example. In each of these situations, we are trying to decrease the likelihood that a person will do the same thing again. There is also "extinction", which is ignoring an undesirable behaviour (not reinforcing) in order to reduce its likelihood of continuing in the future. So when Jett barks and I don't want her to, my best bet is to ignore, rather than punish, the behaviour so it will be extinguished.

Positive feedback is often absent at work, in sport, in relationships, and elsewhere. Positive feedback is a form of reinforcement in that we receive good, positive words of encouragement to support our good habits, choices, actions, thoughts, and decisions. It increases the likelihood that we and our brains will continue to do these things.

Even after studying so much psychology, I didn't fully get what positive feedback was until I did my yoga teacher training. I misunderstood the concept and undervalued the benefits until I experienced it in practice. During my training, each of the teachers-to-be were required to do a practice-teaching session. I couldn't help but notice the thoughts going through my head as I witnessed everyone teach (including myself). "This should be better." "That wasn't very good." "Must learn how to do that." "Should try this next time." "OMG this is awful!" and "Terrible teacher!" But after each practice class we each had to offer our peer teacher-in-training one positive comment about what we felt he or she did well in the demo class, and the feedback session ended with the teacher stating his or her positive comments. I struggled through most of these sessions, wanting desperately to say what I thought was wrong. I felt so confined by positive comments when so many negatives surfaced in my mind. My filters were set. Finally, by the time all 15 of us had presented ourselves in front of the group to be critiqued, I understood what we were doing. I understood, for the very first time, that positive feedback was necessary as new skills were being developed. We need to know what we are doing well in order to continue those behaviours. We need to increase the likelihood that those behaviours will persist even while learning to manage unwanted behaviours.

The funny part about my learning this was that not only did I not understand positive feedback, but I also thought that the fact that I was giving advice on how to be better *was* positive feedback. I thought that the fact that I was helping people to improve and to get rid of the negative things about their teaching, for the sake of being better, was positive. I thought that every time I wanted to change myself into something better that I was focusing on the positive. Even when people around me frequently asked for positive feedback, I still didn't get it. I thought I was delivering it.

Effectively delivering positive feedback is so incredibly important and can be easily overlooked. Many of us have grown up in a culture of "needs improvement." Although, this is possibly not true of many Millennials, who may have come to expect positive feedback and rewards frequently. But among a huge segment of the population, positive feedback is overlooked. As a captain of my ultimate team I now work hard to bring in a sense of positive feedback and encouragement. Naturally, we also focus on all of the things going wrong and want to teach people how to stop doing negative things. But in order for those unwanted behaviours to filter out, we need to focus on strengthening the positive behaviours. When faced with uncertain circumstances, hearing "YES! Keep doing that. It's wonderful!!" is very helpful. It provides stability, confidence, and a foundation to improve in other areas. Hearing too much of "NO! Stop that. Don't do that. That's wrong!" makes for shaky ground where people wind up feeling confused about what they should be doing, without a leg to stand on. Offering constructive criticism is not bad, it just can't replace positive feedback and, in fact, should be kept to a level below the positive feedback offerings. Focusing on positive builds confidence and leads to success by increasing the likelihood of wanted behaviours. The negative behaviours can then be managed either through ignoring them (extinguishing them) or negatively reinforcing them, not punishing them.

So I suggest the often-used sandwich approach: 3-5 positives for every criticism with the criticism in the middle. Invoking that approach with others resulted in the unexpected benefit of changing my own mind. Instead of seeing the negative in people — particularly those who irritate me — when I focus really hard on their good qualities and what I do like about them, my brain and mind literally begin to change and those qualities become more pronounced. What was once a skewed version of how other people needed to change has become more of an appreciation of what people *do* bring to the table. I changed instead. The more I am able to appreciate people and their wonderful idiosyncrasies, the better my life feels, and the more I am able to feel love and compassion toward those around me no matter what qualities they possess. And those feelings are preferable to walking around thinking what a jerk everyone is!

Journal or Track Gratitude

I walked into the first session of the Canadian Positive Psychology Association meeting to find postcards distributed on each table. I picked one up and saw that they were "appreciation cards," and on the back was space for participants to write one point of gratitude or insight gained as a result of the session. These cards were collected at the end and offered to the facilitator as a source of positive feedback. It was a perfect example of positive psychology in action, showing that gratitude makes an impact on the receiver and the giver. I was so excited to fill it out when I was a participant in a session. It allowed me to focus on what was really good about the session. I started to feel my filters shift as I became more and more aware of all of the good things happening at this conference. In retrospect, this is probably one of the best marketing tactics around, and it probably worked well for them. I attended their second conference ever, and the attendance doubled from the first to the second conference. I suspect with so many glowing reviews that they will double again next time! I was also excited to know that I would get a whole pile for myself as a facilitator the following day. I was reminded of what people like about my sessions and learned some new things as well, and that increases the likelihood that I will continue those behaviours!

Focus on Strengths — Your Own and Others'

Later in the conference, I heard several scientists discussing the value of focusing on strengths over weaknesses from a scientific perspective. The science on the matter says that we are happier when we are doing things we are good at, and less happy when we are improving our weaknesses. This concept is inherent in positive psychology.

When trying to apply this to myself, I realized I often pay little attention to my strengths. Why would I focus on them when there are so many things I need to make better about myself? As an athlete, I grew up this way — being told how to improve. But that isn't as effective as being told what you're good at. Building on strengths allows us to flourish and expand those good qualities that we often take for granted. Developing our strengths allows us to have a supportive foundation. It's a manner of increasing the likelihood of certain behaviours, namely our strong characteristics. When we are confronted with challenges or difficulties, our weaknesses are not going to help us out, it's our strengths, by the very nature of their name! We should rely more on our strengths!

Strengths Exercises

Each week list some of the tasks that you achieved. For each, describe a strength that allowed you to achieve that task. Collect these strengths over several weeks and look to see if there are patterns or common themes among the strengths that emerge.

Ask someone for feedback on what they think your strengths are. You could ask a personal friend or a professional colleague, depending on what you are seeking to learn about yourself.

Notes:

Focus on Solutions

Similar to focusing on our strengths, solution-focused therapy offers a solution approach focused on determining what is going well, what successes have already been made, and what strengths were present in those successes. For example, imagine we are in a very difficult working environment. We are feeling stressed and are lacking rewards. Then we are asked how we feel about our working environment, on a scale from zero to ten, ten being wonderful and perfect, and zero being terrible and awful. We rate our workplace a four. One approach — the problem approach — would have us dig up all of the issues that are keeping us at a four. We may come up with some ways of mitigating these negative circumstances, but more likely we walk away from the conversation feeling defeated, awful, and generally negative in both mind and body. But imagine the alternative when we adopt a solution approach. We are asked to reflect deeper on what it is about our situation (or ourselves) that allows us to claim a four and not a zero, for example. We begin to consider places where we feel in control, places of reward (no matter how small), and the many ways that we are not at a zero. That instills confidence that can be leveraged. It becomes the foundation of strength. Then we begin to focus on how we can incorporate more of those elements, more of our strengths. We end up coming up with a solution that leverages our strengths rather than makes us struggle against our weaknesses. Brilliant actually!

Solution-Focused Exercise

Consider something you think you cannot do, but instead of thinking of the task as all or nothing, consider it as a likelihood of events. Consider how likely you are to be able to achieve the task on a scale from zero to ten (ten being that you most likely *can*). If you choose any number higher than zero, delve in to understand why you picked that number. Ask yourself what strengths, resources, or abilities allowed you to see that number and not the number below it. Focus on what you *can* do, *do* have, or have *already* achieved rather than how far you have to go.

Notes:

Intellectually, many of us recognize the benefit of staying positive and perhaps we were further convinced by this chapter. We may also recognize the good intentions of someone who tells us to "be positive." But "being positive" does not actual describe what we are feeling, as it implies. It is a descriptor of what we should feel. But with every "should," there should be caution. Being positive is something to strive for, as a reminder not to linger in a negative moment unnecessarily. But it's not something to pretend to be. It should never replace what is happening in the moment. We cannot simply convince or attempt to convince ourselves that we are positive. We have to genuinely feel the sense of positivity inside us or provide enough evidence to let it come through on its own time, with the goal of genuinely becoming positive. In order to feel anything positive, genuinely, we might have to work really hard at altering our filters and priming ourselves to be positive. And this is where the practice of positive thinking comes in. We might even need positivity mirrored back at us, if we truly don't feel able to access it. In any case, there are several ways we can change our brains and learn to think more positively and optimistically and enjoy all of the positive physical and mental health benefits associated with such a demeanour.

Chapter Assignment:
Thought Reframing

The purpose of this assignment is to learn to transform a recurring, automatic, and/or negative thought that holds you back into a thought that allows for action and a forward movement.

Use the steps provided below to familiarize yourself with the process of thought reframing.

Step 1: Identify a negative thought.

Probe to come up with a thought you might have regularly, like the example provided below in the chart.

Step 2: Inquire about the thought.

Ask yourself if the thought is exaggerated in any way. Look for keywords like "always" or "never," whether explicitly stated or implied. Consider the three Ps of pessimistic thinking (Personal, Permanent, and Pervasive) and identify any of those elements in your thinking.

Step 3: Identify how the thought makes you feel.

Ask yourself how the thought makes you feel. Consider bodily sensations as well as emotional words, as you did in the exercises earlier in this chapter. Dig deep to really understand how it makes you feel.

Step 4: Ask if the thought is a hindrance or help.

Identify whether the thought is hindering. It likely is. It may serve some kind of a role, but ultimately, seek to identify how it hinders you.

Step 5: Reframe the thought.

This is the tough part. Come up with a new, but true, thought. Try to shift it into a non-personal, non-permanent, and non-pervasive thought to avoid the three Ps of pessimism. The new thought must feel true to you. Ensure you feel good about this new thought and believe it. Realize that you may have to work very simply at the start. Also, identify the action involved in the thought, as opposed to the inaction from before. You could also discuss any solutions to the situation that come up as a result of this new thinking.

Repeat with two or three more thoughts until you start to see how your thinking pattern could be holding you back from coming up with solutions and actions. This will help you feel more in control.

Notes:

Chapter Assignment:
Positive Feedback

Offering positive feedback at any time is great. Try any or all of these exercises.

Offer unsolicited positive feedback to someone you like.

Offer unsolicited positive feedback to someone toward whom you feel neutral.

Offer unsolicited positive feedback to someone who challenges you.

In a group, prepare one positive comment each for all other members of the group. Collect all of those positive comments anonymously in an envelope. Give the appropriate envelope to each person and let them read all the positive comments. It's a great way of developing group cohesiveness and team bonding. I do this with my sports teams and coaching classes!

Notes:

Chapter Assignment:
Gratitude Exercises

Keep a gratitude journal. Each night before going to sleep, record three things that you're grateful for.

Send thank-you notes to people. They are wonderful to receive, and also wonderful to write.

Send holiday cards out at Thanksgiving!

Notes:

Chapter 6

Success through Failure

If I asked you to sit back, reflect on some of your biggest failures, and then tell other people about them, you would probably start to feel uncomfortable. If I started to list my own failures, we *both* might start to feel uncomfortable. Failure makes most of us feel uneasy even when we observe it in others because it mirrors to us what we try so desperately to hide — our own inadequacy. I remember starting to shift in my seat when I first heard Toronto medical doctor Brian Goldman, host of the CBC's *White Coat, Black Art,* speak about his failures in his TEDx Talk.[38] He started with a story about being a resident in emergency; he let one of his patients go home without doing the right checks, and that patient later died. Then he told stories of medical error one after another, and I remember thinking, "I can't believe he's telling us all of this!" I felt equally uncomfortable a few years later when I heard him again speak of failures during a keynote address at a conference devoted to failure, a conference

---

38. Brian Goldman, "Doctors Make Mistakes. Can We Talk About That?" *TED Talk* (November 2011),
http://www.ted.com/talks/brian_goldman_doctors_make_mistakes_can_we_t alk_about_that

where I too was speaking and was divulging my own failures. Being wrong, however we word it — failing, making mistakes, committing errors — makes us feel uncomfortable. It threatens our identity, and without adequate confidence and understanding, failure can paralyze us.

Despite our negative relationship with failure, failure is actually an incredible learning opportunity, psychologically, emotionally, and neurologically. When I think about my major accomplishments, I often attribute my successes to having learned how to fail in sports. Without the inevitable losses that come with sports competition, I can't imagine how I ever would have learned how to win.

Accepting our failures as learning opportunities is challenging and requires us to acknowledge them outright. Let me start: I have many failures under my belt, but here is one of my favourite and worst mistakes from my PhD days. It happened when I was copying and pasting some of my scientific data from one file to another. I made a mistake and copied the results into the wrong headings without realizing it, essentially giving the wrong group of animals credit for an effect of treatment. My supervisor and I began to interpret this false data along with other pieces of information, and we developed a really nice story to tell about potential mechanisms of action for anxiety in rats. We were preparing to write up the results to be submitted to a scientific journal. Then, while double-checking all of my numbers (thank goodness I do that!), I realized I had made this huge mistake that entirely reversed the results we thought we had. I was devastated, stunned, and absolutely defeated. I sat motionless in shock for at least an hour. I didn't know what to do or how to overcome this huge mistake. I kept it to myself for about 24 hours, thinking through all possible scenarios, including my supervisor kicking me out of her lab for being such an idiot. I started thinking, "How could I ever be a scientist if I made errors like this?" So many words of shame went through my mind. I considered just leaving without a word. Crawling under a rock and hiding forever seemed way easier than having to fess up. Eventually, I became more rational and knew that I had to come clean with my supervisor and suffer whatever wrath was to befall me. The 24 hours spent worrying nearly killed me.

Thinking of that big mistake made me feel sick to my stomach for quite some time afterwards, but I got over it, and it's worth sharing that everything worked out in the end. I didn't quit and I didn't get fired. My supervisor was okay with the new result and even the fact that I had made a mistake. Her words to me weren't nearly as harmful as the words I had used for myself. And the data survived to be published with the correct interpretation.

In most circumstances failures hurt, and we naturally try to avoid them. That strategy might, however, be counterintuitive. Despite the serious discomfort of failures, making mistakes actually serves to increase success in the future. Failures and mistakes are natural, and they serve a neurobiological process inherent in each of us. They serve as important feedback mechanisms to direct and correct future behaviour. Essentially, they lend themselves to a learning process that improves our performance in the future. Without failures we would not progress. Do you think the Wright brothers got flying right the first time? Did you know what Michael Jordan didn't make his school basketball team his first time trying out? Did you know that Lucille Ball was considered a "failed" actress before her role in *I Love Lucy*? Did you know that Steven Spielberg was rejected for admission to film school at the University of Southern California? There are many examples of people we think of as highly accomplished having failed big. In fact, probably everyone who has succeeded had failed at something beforehand.

We don't have to look to the stars to see examples of success through failure. A successful colleague may have applied to 10 different companies before landing the job you see her in now. Your boss may have been reprimanded for failing to make quotas. Your now-amazing assistant may have been fired before he landed this job with you. What successful people who have failed all have in common is that they didn't give up. A coaching colleague of mine, Dana Warren in Newfoundland, recently wrote a blog post on the subject, and in it she says "if you fall down 7 times, get up 8." Well said. That reminds me of Jack Cantwell, author of *Chicken Soup for the Soul* (and many other titles). He claims to have submitted his manuscript over 100 times before it was accepted. Those agents and publishers who failed to see his value as an author probably learned from *their* mistakes too!

Mistakes will happen. It's how we deal with them that matters. Fortunately, the brain has a very strong way of using the mistakes as feedback, adjusting for the future, rebounding from mistakes, and ultimately setting us up for future successes.

The Detection and Adjustment System in the Brain

As discussed earlier, all of our experiences leave footprints on the brain, and the brain is constantly updating and adjusting to those experiences. Failures are no different in that they are simple learning experiences like all other experiences. Making use of error for future success happens through a simple detection and adjustment system in the brain. This can be observed by looking at the brain's electrical activity in response to correct versus incorrect responses when given a specific task. The task simply involves asking people to respond to which side of the screen a dot shows up on and then indicating the answer with a stroke of a key on the keyboard. Clear differences in the brainwaves appear when people make correct versus incorrect responses.

These brainwave spikes are called "error-related negativity," and appear within milliseconds of a person realizing they have made a mistake. The spikes can vary in terms of size (greater or lesser *amplitude*) and are predictive of the extent the mistake influences future success. For example, researchers have noticed that when the spike is big, people tend to do better on subsequent trials compared to when it is small or not present at all. As a result, this spike has become known as the brain's error detection-and-adjustment system. In other words, the brain activity associated with the spike detects mistakes, and then recommends an adjustment for the future, an adjustment that ultimately lends itself to success.

There are two areas of the brain that seem particularly important in generating this detection-and-adjustment spike. The first area is the *medial prefrontal cortex* (MPCx). The MPCx is involved in behaviours like self-monitoring. For a detection and adjustment system to work, the brain needs to keep track of how we are doing in the task at hand. The MPCx seems to be the area involved in that. The second area is known as the *cingulate cortex*, which is involved in other behaviours related to detection, including general awareness and keeping a goal in mind. But it also seems to be involved in the comparison of the goal and the person's alignment with that goal. For example, if our goal is to make correct responses, the cingulate cortex seems to be the area of the brain that keeps track of how well we are doing with that goal. It's like having an internal coach creating accountability for us around the achievement of our goal and, ultimately, the success.

Knowing exactly what other brain areas are involved in this detection-and-adjustment system is still under scientific speculation. But what is clear is that the brain does respond to failures and it makes adjustments for future behaviours, based on monitoring and awareness. In fact, if we broke it down enough we would find that the principles of trial and error are absolutely inherent in the entire manner the brain works. We are born with so much to learn as we move out of the comfort of the womb and into the very unpredictable world. Even our very first steps are met with significant trial and error. I don't know of one person who learned to walk without ever falling, and I know of no one who didn't get up after that first failed attempt. We are very accepting of a child's trial-and-error process, yet at some point we expect ourselves and our fellow humans to be without error. That type of thinking, whether overt or covert, is absolutely flawed, at least from a neuroscience perspective. Our brain was designed (or has evolved) to learn, and with every piece of learning come failures. But our brains are inherently resilient, or neuroplastic, and as a result we adapt to each failed attempt with an adjustment for future attempts. The only way the brain truly fails is if we never try again.

Not only is the brain wired to deal with failures and mistakes, it is bound to do it time and time again. It's set up that way. Denying this truth only puts us in the unfortunate situation of striving for perfection, which does not exist. The expectation that we might go through life without failures can set us up to be chronically disappointed. Neuroscience can teach us that not only is being imperfect natural, it is inevitable.

Failure Paralyzes Us

While talking with Annie, a beautiful soul and wonderful life coach, about the very nature of rejection, she said to me, "Rejection in others is inspiring. You never get inspired by the people who do nothing. You get inspired by the people who try even if they fail." I had never thought about it that way, but it rang true immediately.

Sadly, some people don't try. My friend and colleague Allison was working with a client on helping him find a new career. There was a job posting that the client found and thought sounded great. When Allison asked her client if he was going to apply, the client abruptly replied "Absolutely not! I never apply for anything unless I know for sure I will get it." When Allison told me the story, I was dumbfounded. It had actually never occurred to me that someone would not apply for things without 100 per cent certainty of success. My personal process is much different. I apply to things left, right, and centre. Once I even rewrote a job ad, telling the search committee in my cover letter that I would be well suited for the job if only they required these skills! True story![39]

As I began to ask around I realized that this client was not an anomaly. In fact, many people are so paralyzed by the fear of failure that they avoid any chance of it, effectively rendering themselves unsuccessful simply because they don't try. Although comfortable in the short term, how unfortunate for them in the long term.

If I'm honest, I too have been paralyzed by failure. By the time I was doing my post-doctoral fellowship, several scientific papers had been published, and never once had any been rejected. Of course, I was proud of this, knowing that publishing can be quite difficult. But I happen to come from a lab from which papers were always accepted. So when I got to my post-doctoral training, I had my first manuscript rejected. I was outraged and, admittedly, I didn't know how to handle it. I thought the paper was excellent and knew several researchers in the field who did too, yet it was rejected and I took it personally. Unfortunately, this rejection really did cause me to fail. I never resubmitted that paper, and I have not submitted an academic paper since! Now, when I look back, I can see very clearly that I actually failed because I never tried again. A part of me feels sad that I didn't have the resiliency, the mentorship, the awareness, or the perseverance to try again. A big part of me wonders "what if." In hindsight, that is the failure I am truly most ashamed of, only because I failed to try again.

---

39. I didn't get the job or an interview, by the way, so I would not recommend this strategy.

That feels worse than the initial sting of the rejection. I almost went down that path with this book actually, but I'm glad I didn't.

Neurochemical Changes Associated with Defeat[40]

It feels damned uncomfortable and our egos take a huge blow when we fail. One way to see it is "social defeat," which is one way we look at failure when we study the neurochemical changes associated with it. One of those chemical changes is a drop in testosterone. Testosterone is a hormone present in both males and females, but we often think of it as being only a male hormone. Testosterone levels have been associated with social dominance, typically studied in males. In the laboratory, we can measure failure when we place two animals that normally exist within a social dominance hierarchy together. Rats, for example, live in a social hierarchy in which one male rat is at the top as the "alpha" male. He would likely have fought his way to the top, or perhaps he was simply the biggest of the bunch and no one dared pick a fight with him. Once at the top, he rarely has to fight to maintain this status. The rest of the rats are organized in ranks underneath, for example, the next in line are the "beta" rats and so forth.

Once the hierarchy is established, it is relatively stable until something changes, perhaps an external stressor or a new member to the colony. If, however, two rats of equal size are placed together in a neutral environment, they

---

40. Brenda F. Reader et al., "Peripheral and Central Effects of Repeated Social Defeat Stress: Monocyte Trafficking, Microglial Activation, and Anxiety," *Neuroscience* 289 (2015): 429–42.

will exhibit a variety of typical behaviours in an attempt to prove which is the more dominant and, therefore, which is the "alpha male." The losing rat adopts a subordinate status and can be relatively happy to coexist as such. After the initial shock of losing, having a stable hierarchy is less stressful than fighting regularly to maintain a high status. This laboratory paradigm has been used extensively to discover how social defeat leads to a variety of depression-like symptoms. Serotonin is another neurochemical that changes in line with social defeat. Similar to testosterone, serotonin drops with a loss and rises with a win.

These chemical changes are best studied within a social dominance model like the one just described. Other species, including mice, chimpanzees, and lobsters, all have social dominance hierarchies, and the same pattern of changes has been observed. When an individual loses, testosterone and serotonin levels both go down in that individual. The levels of hormones and chemicals are state-dependent, meaning that they change based on the rank of an animal (including a human) within a hierarchy. For example, in one study, when a dominant male was removed from its group, the next-in-line male that replaced him showed an increase of approximately 60 per cent in blood serotonin levels.[41] We also know that social defeat increases two stress-related hormones: adrenocorticotrophic hormone (a hormone released by the brain as part of the HPA axis referred to earlier in the chapter on stress) and corticosterone (the non-human

41. Michael J. Raleigh et al., "Social and Environmental Influences on Blood Serotonin Concentrations in Monkeys," *Archives of General Psychiatry,* 41, no. 4 (1984): 405–10.

version of cortisol, the hormone released by the adrenal glands in response to the first surge). Such defeat not only prolongs the HPA axis but also produces alterations in the serotonin system.

Humans are not exempt from these chemical changes. However, most of the direct research examining this relationship in humans has been done using sports scenarios. Winners of tennis matches and soccer games have shown this surge of testosterone and serotonin while losers show the opposite. Possibly more interesting, even spectators of a sport have been shown to exhibit a similar pattern of neurochemical changes when their team wins or loses!

Social defeat is a good paradigm for considering the effects of failure on the brain and body. Although most of the science deals with animal hierarchies or human sports, we can consider other scenarios from our lives, for example, when we feel personally attacked. If someone yells at me or tells me I am wrong, or if I lose an argument, I can feel it. If I am excluded from a social group or not invited to a party, I can feel it. All those rejection letters I got? Those were essentially social defeat too. Even more interesting, social defeat can be self-generated and is highly dependent on our self-perceptions. I can also feel defeat against my own inner critic when it tells me that I can't do something, when I get confused or twisted in my mind as I prepare to write something (like this), or when my inner critic tells me I am not good enough in any respect. According to a review article by Taylor and colleagues,[42] defeat can result after: a) a failure to attain, or the loss of valued resources, including social and material resources; b) social put-downs or attacks from others; and c) internal sources of attack, such as self-criticism, unfavourable social comparisons, or unachievable ambitions. Each of these scenarios is likely accompanied by a drop in testosterone and serotonin and an increase in stress hormones.

Self-Reflection Exercise

Think back to an experience you've had that would be considered "social defeat." What did it feel like? Did you feel any of the sensations of stress? What about a social victory? Did you feel strong? Did you feel a surge of testosterone? Did you feel the goodness associated with an increase in serotonin?

Notes:

---

42. Taylor, PJ, Gooding, P, Wood, AM, & Tarrier, N. (2011). The role of defeat and entrapment in depression, anxiety, and suicide. *Psychological Bulletin*, 137 (3).

Failure Is Stressful

Failure causes a release of stress hormones. These hormones hit us both psychologically and physically. This is another reason we try to avoid failing. All of those uncomfortable feelings are enough to make us avoid situations where we are liable to fail. However, avoidant behaviour can lead us to do, think, and feel many different things. For example, in attempts to avoid confronting failure, we may self-sabotage. In other words, we might stop ourselves from failing — and succeeding — before we get a chance to do either. An example might be going out and getting really intoxicated the night before a big test so that if we fail, we can blame it on the drinking. Or not renewing a gym membership so that we can blame our lack of weight loss on the gym that is too expensive. Or taking on too many paying contracts and using that as an excuse not to write this book. Self-sabotaging happens so that we feel that our failures are within our control. We achieve that feeling of control by blaming the failures on something else. It's a good strategy if the goal is to protect our egos. It's not a good strategy if we ultimately want to succeed.

We can also set ourselves up to avoid these uncomfortable, stressful experiences of failure through rationalization, a defence mechanism. "Oh it wouldn't be worth it for me to go back to university because it's too expensive," or "I can't take that art class at the community centre because I have to be available in case a job comes up." We can also adopt a variety of other strategies like humour, anger, blame, concealing, justifying, defensiveness, and apathy, all of which cover up the sadness and a stressful experience of failures that we don't want to feel psychologically or physically.

I remember how my mind and body dealt with receiving some negative teaching evaluations. First of all, I barely noticed the good comments and honed in on the negative ones immediately. As I read them alone to myself, I got really angry. The voice in my head yelled in defence: "You don't know what it's like to put up with all of your crap and all of your demands every day! You don't know how hard I try to give you a great learning experience! You don't know all of the things that I do all day long just to help YOU!" Of course, the voice in my head was trying to protect my ego. But if our goal is to actually succeed through failures by learning from our mistakes, this strategy holds little value. When I managed to settle down and distance myself (my ego) from the reviews, I was able to piece them together and understand that from the students' perspectives, in this context, I was not fulfilling their needs. More important, when I was able to calm down and take the information in, I was also able to figure out how to adjust my next course delivery to account for those unmet expectations and how to provide a better learning experience for my students. If I rested in the place of blame, anger, and defensiveness, I would do lots to support my ego but little to support my growth as an educator.

Letting the Sensation of Failure Pass By

The physical experiences of failure (testosterone drop, serotonin drop, stress hormone elevation) are all very real and unwanted experiences. But they don't last forever and they can be experienced rather than avoided. We can sit with the experience, feel the sensations, tap into what the mind, body, and brain are doing, and watch the sensations pass by on their own time. Remembering nothing lasts forever or that everything changes is a helpful piece of ancient wisdom we can borrow from yoga philosophy and mindfulness. The space between those physical experiences rising and falling and our automatic mental reactions to avoid them is our playground for change, and it requires simply awareness. First it's important to realize that there IS a space there; there is a gap between the physical experiences and the mental experiences. Our physical reactions to sensation are not the same as our mind's reactions. We don't have to invoke a defence mechanism. We can employ better brain power: consciousness. Second, realize that the physical doesn't have to give rise to the mental. The physical experiences of failures and mistakes are very real and may never go away. In fact, we want them to occur so we are able to detect when something has gone wrong and adjust, as the brain is primed to do. If we never knew we failed, made a mistake, or caused an error, how would we ever learn? We wouldn't. So those physical sensations are simply signs that something went wrong. We never want those sensations to go away entirely. But we do want to be able to better manage our reactions to those sensations.

Success Through Failure

Fear of failure simply must be overcome, and my hope is that the perception of failure as a natural neurobiological process that serves us (rather than hinders us) will help us get past the fear. Long before we ever considered failure a shot to the ego, we each used failure to try and try again. How else did most of us start walking? Or talking, making grammatical errors as we began to learn the illogical rules of English. Or manners. Or mathematics. Or spelling. Or colours.

Several strategies can be used to overcome the fear of failure, which often paralyzes us, and for dealing with the discomfort of failures as they are happening. For example, cognitive–behavioural techniques can be used to help modify appraisals and cognitively restructure the situation so as to reduce the individual's sensitivity to signals of defeat.[43,44] We have used this strategy before when we covered optimism and happiness. The same principle applies. We can learn to argue against ourselves, see that our failures are actually successes, and notice what we have learned as a result of that failed attempt. Also guided reimagining of past defeating experiences could be used to shift our thinking around these events.[45] We can build an image of a more positive and dominant identity for ourselves by emphasizing the ways we have demonstrated resilience in the face of these experiences, and by highlighting our other successes.[46] Finally, and this is probably my favourite in terms of confidence building, is physically putting our bodies into the dominant

---

43. Judith Johnson, Patricia Gooding, and Nicholas Tarrier, "Suicide Risk in Schizophrenia: Explanatory Models and Clinical Implications," *Psychology and Psychotherapy: Theory, Research and Practice* 81, no.1 (2008): 55–77.

44. Stephen R. Swallow, "A Cognitive Behavioral Perspective on the Involuntary Defeat Strategy," in Leon Sloman and Paul Gilbert (Eds.), *Subordination and Defeat: An Evolutionary Approach to Mood Disorders and Their Therapy* (Mahwah, NJ: Erlbaum, 2000), 181–198.

45. Deborah Lee, "Case Conceptualisation in Complex PTSD: Integrating Theory with Practice. In Nicholas Tarrier (Ed.), Case *Formulation in Cognitive Behaviour Therapy: The Treatment of Challenging and Complex Cases* (London, England: Routledge, 2006), 142–166.

46. Nicholas Tarrier, "Broad Minded Affective Coping (BMAC): A 'positive' CBT Approach to Facilitating Positive Emotions," *International Journal of Cognitive Therapy* 3 (2010): 64–76.

positions that animals adopt. When two animals get together they often start to puff up, trying to look as big as possible. Ideally, they wouldn't have to fight and could win based on size and confidence alone. I bet we could do the same if we opened up our chests to the world and paraded around like we were strong, confident, and dominant. "Walk the walk," so to speak. Sure, we might overdo it, but a little extra confidence with good intentions could probably serve many of us. I actually know a woman who, when having to make difficult phone calls (and when in the privacy of her own home), will strap on a fake penis to help her feel more masculine in the conversation. She is convinced that feeling more masculine allows her to be more assertive in her conversation. If I set my feminist ideology aside, I can see this as an interesting mind-body connection strategy!

Examining Our Relationship with Failure

Here are some formal questions to guide you through a process of understanding your relationship with failures, acknowledging your failures, developing a healthier and more productive relationship with failure, and feeling the fear and doing it anyway!

My Relationship with Failure

I doubt any of us is a stranger to failure. I know I'm not. I failed at school a *lot* before university. I was a "dumb kid." But somehow I went from dumb to doctor, and I really don't think it was because I was anything special, a gifted child who had not been challenged in school. I think a big reason why I am successful in many areas of my life now is because I am not afraid of failing. If I were, I wouldn't have written this book, for example. This is not to say I like failure — because I don't. I hate it. It's as uncomfortable to me as it is to you. But I have learned that it is part of life and I embrace it in my life.

A few years ago, I received a rejection from Team Canada for ultimate Frisbee, which I have devoted more than a decade of my life to. Although not making the team didn't come as a big surprise to me, it still felt the way a rejection notoriously feels — awful. I felt a desire to cry as sadness rushed in. I felt pressure behind my eyes. Then later I felt sick to my stomach, a ball of knots, a ball of defeat. In my heart, I felt unworthy and not good enough. My dream had been crushed, and I felt like I had lost purpose, at least temporarily, because making the team was something I had desperately wanted.

I have earned rejections from universities, including prestigious ones like the University of Toronto. I have been rejected for scholarships and jobs, by granting agencies and other sports teams, and by romantic partners and friends. When I was an undergrad, I applied to seven schools for my graduate studies. One by one, the rejection letters rolled in, five in total. I also applied for a fellowship with the prestigious Natural Sciences and Engineering Council (NSERC) of Canada four different times and was rejected each time.

I applied for student travel awards to present my research at conferences. One in particular was specifically for women in science. I had to write an essay about the barriers for women in science. I researched many different angles and drew upon my own personal experience and from mentorship programs that I had been involved in. In exchange for this hard work, I got rejection letters time and time again. When I started to apply for research grants to support my studies as a postdoctoral fellow, my sadness and defeat eventually turned into ammunition and drive.

I have also been rejected in love. One time in particular was devastating because it was my first attempt at a relationship after ending a 12-year relationship with my high school sweetheart. I was 25 and had fallen quickly for someone new, but it barely lasted a month before I was rejected. It hurt. A lot.

With all of this rejection, I should feel like a big loser, devastated, worthless, and paralyzed. But I don't. Why not? Because with the pile of rejections have come so many opportunities. Along with the five rejection letters I received from grad schools, I received two acceptance letters, one of which I accepted and one *I* rejected! Three years later, when I applied to do my doctorate, I received three offers for my three applications. Success x 3! And although my first four NSERC rejections took a bite out of me, the next two were well received and I ended up with funding for two years of my PhD and two years of my post-doctoral training. Meanwhile, along with all of the rejections for the many travel and research grants I applied for came an equal number of cheques and acceptance letters. Being rejected in love turned out to be an amazing experience too. That person told me that my emotions were beautiful and part of what that person loved about me. We are still friends today, and that rejection was, in hindsight, one of my favourite and most useful. The feedback allowed me to enter relationships openly, sharing my inner self and wearing my heart on my sleeve, not fearing that those emotions might push someone away. As for friends who rejected me? I guess, you win some and lose some. Many of us dream of pruning our friend lists down to only those with whom we feel mutually supported. How can I fault people for that? In fact, I preach it regularly!

Finally, despite getting a rejection from Team Canada, it serves me well to remember that, prior to that, I received two invitations to try out for a team that would go on to be one of the best in the world. A few years later I ended up with some of those very same talented players as teammates when we won a gold medal at the Canadian Nationals in 2014. If I had never tried, I would never have had the confidence to play in that circle. So I guess I was good enough after all, once I changed the scope of my vision. Oddly enough, the feedback that I received upon not making the team was that I could have had a larger impact at tryouts had I been in Toronto a year earlier, playing and learning their systems. What stings about this is that I actually could have been in Toronto a year earlier, but I hesitated in coming partially because I was afraid to give up the comfort of Halifax, including the team I was playing for there. I was too afraid to move to the big city and attempt to make my way into the ultimate scene (and the professional scene). I was afraid of failing. The irony of my Team Canada rejection serves as a good lesson.

I didn't get all of these rejections because I was an idiot, a loser, or damaged, just as I didn't get all the opportunities, awards, scholarships, positions, and publications because I was anything special. I got all of these because I was not paralyzed by the fear of rejection and because I knew that the risk of applying was merely a blow to my ego, and I figured I could handle a few dents.

Rejection stings, indeed. I probably felt a drop in testosterone and serotonin and a rise in stress hormones that corresponded with those stings. Failure can be devastating and crushing and nauseating. Sometimes it brings us to tears and makes us feel hopeless and helpless and pointless. Sometimes it makes us feel utterly worthless. Sometimes it defeats us and brings us to our knees. Sometimes it makes us run away and hide. Sometimes it brings us shame. Sometimes it even kills us. These can all be true experiences. But not one of those rejections ever hurt me as much as the feeling of lost opportunities, regret, "what could have been," uncertainty, or as much as the smallness I would feel if I had never tried in the first place. Ultimately, failures have allowed me to do well in school, win gold medals in sports, start my own company, write this book, secure funding, get pregnant, and fall in love. And I couldn't be any happier about them.

Self-Reflection Exercise

Most embarrassing failure:

Most embarrassing mistake:

Greatest lesson learned from a failure or mistake:

Failure or mistake that causes you the most regret and why:

Biggest mistake that you never fessed up about:

Failure that hurt the most:

Mistake that hurt the most:

Favourite failure:

When you think about all of these failures and mistakes above, how do they make you feel?

> In your body:

➢ To your ego:

➢ In your heart:

Mark an x where you fall on each spectrum below.

I hate failure	....	I love failure
I cringe about failure	....	I thrive on failure
I accept failure	....	I try to avoid failure
I learn from failure	....	I ignore failure
I admire people's failures	....	I judge people who fail
I try to hide my failures	....	I am open about failures

Can we sit with the discomfort of the physical experiences that accompany failure and not try to change them or make them go away? Can we sit with the discomfort of the physical experiences and watch how they change over time? If we can, we will probably learn that the sensations dwindle over time, even without any mental reactions such as anger, apathy, or humour. Over time, those sensations will subside and the failures won't feel as bad again. Essentially, we are going through desensitization training in attempts to develop better relationships with our failures.

Avoiding Unnecessary Mistakes

Despite all of this knowledge of imperfection being natural and inevitable, most of us still seek to avoid it. We can't prevent failure, but we can make decisions more consciously in order to lower the likelihood of such mistakes. In circumstances where accuracy is important, we can bring awareness and consciousness into our experience and into what we are doing. By doing so we move out of a realm of automatic thinking, which is prone to errors because automatic thinking inherently favours speed over accuracy. Instead, we bring in an element of consciousness or "controlled thinking." Controlled thinking requires more time and patience, and favours accuracy over speed. There is always a trade-off when it comes to how our minds and brains work, and we have to decide if we want to favour quickness or correctness. As a result, instead of relying on our quick heuristics, shortcuts, and automatic thinking, we have to think more intentionally (with control) about what we are doing. It is indeed possible and will result in fewer errors. Knowing when to bring in awareness and when to work automatically is helpful for ultimate efficiency! In fact, we did this earlier when we talked about reframing our negative thoughts.

When pressure is high, mistakes happen more frequently. In lab experiments, we put the pressure on by asking people to do a task "as quickly as possible." This often leads to decreased performance and accuracy. We can also increase mistakes by instituting punishments or negative consequences when someone gets something wrong. We can think of situations where that happens in real-life environments like work, sport, or even relationships and parenting.

Mistakes also happen by virtue of our brains working to be faster and automatic. Not having to think about something takes less effort and uses fewer resources. When things are new, the brain requires lots of energy. For example, learning how to drive, ride a bike, do simple math, answer phones, greet clients, or anything that has become routine requires effort initially. But there is a trade-off for the automaticity that comes: mistakes! The more automatic something becomes, the more likely we are to make mistakes because we stop paying attention, generally speaking. Here's an example:

Answer this question: If together a baseball and bat cost $110, and the bat costs $100 more than the ball, how much does the ball cost?

You probably came up with a rather quick and automatic response, which was to say that the ball costs $10. But that's a mistake. If the ball costs $10 and the combined total is $110, then the bat would be only $90 more than the bat. The answer is $5 because if the ball is $5 and the combined total is $110, making the bat cost $105, or $100 more than the ball. But most people answer $10 because the brain quickly and automatically subtracts $100 from $110 to come up with $10, which is a mistake.

The brain also likes things to make sense and sometimes it lies to itself and fills in gaps in order to make sense of the world. The visual system of the brain is a good example of this. In certain kinds of damage that occur within the visual cortex (often through stroke), people are unaware that they are experiencing blindness, known as blindsight. The visual cortex (at the back of the head) eventually receives the visual information coming in through the retina of the eye. If the cortex is damaged, it's almost like not having a screen to project on, and the image is lost. When discrete areas are damaged but the surrounding areas are not, a person might receive a partial image. But because the brain knows there *should* be a piece of the image there, it just fills it in. Even though that person may be technically blind to some aspect of the full image, he or she would be completely unaware of it, but technically the image is an error. There are many other similar visual illusions the brain does, which are also errors of perception. Magicians make use of these to trick us! We allow errors to occur so we can preserve our beliefs. This can also happen with self-identity. We might avoid paying attention to pieces of information that inform us, for example, that we are bad managers. Our brains, in defence of our egos, can weight some information less than other information. These biases are known as "confirmation biases." They exist when we seek out and retain information that serves to confirm what we believe because it's easier.

We cannot possibly prevent all these mistakes, nor should we want to. Our brains have evolved to be efficient, and that means mistakes will happen. It's the price we pay for quickness. But in circumstances where accuracy is important, we can bring awareness and consciousness into our experiences. This awareness allows us to rise from our automatic errors and see things more clearly. However, it requires more time and patience, which compromises speed.

During other times, when accuracy is not imperative, trial and error is our opportunity to learn, to fall and rise, and ultimately to succeed. Failure is the price we pay for success. They come together, hand in hand. Understanding the relationship between failure and success is much more complex than simply trying to avoid making mistakes. It's also a more realistic and effective way of existing as the imperfect humans we are.

Chapter Assignment:
Thought Reframing II

Do the same thing with this assignment as you did with the thought-reframing assignment from the previous chapter, focusing specifically on perceived failures (see page 167).

Notes:

Chapter 7

Motivation, Curiosity, and Passion

Our thoughts can get the best of us. We can succumb to negativity, pessimism, and failure or we can direct our thinking into positivity, optimism, and success. We can also direct our attention toward the goals that serve and motivate us. We can tap into our passions and our drive, and we can live more intentionally, feeling empowered by our inner strength and desires. We can tap into basic needs and drives that are both psychologically and neurologically motivated.

All sorts of animals have goals. A rat running through a maze has the goal of finding a sugary treat. A dog sitting patiently has the goal of attaining a treat for doing as told. Humans have all sorts of goals. We aspire to be rich, famous, youthful looking, muscular, or to exercise every day, to lose weight, to drive a fancy car, to be happy, not to be depressed, to be our own boss, to climb Everest, to get into the *Guinness Book of World Records*, to go to the Olympics, to become a rock star... But having a goal is not enough. We need a reason to seek out those goals. We need motivation. Motivation literally means "to move" (from Latin *motivus*, from *movere*, meaning "to move") and it propels us forward in our lives to seek and attain those goals. Motivation and its close cousin "direction" help in achieving a goal.

Motivation can come from within, meaning that motivation is intrinsically generated. Intrinsic motivators include basic needs, food, water, shelter, safety, and air, all elements of survival. Theorist Abraham Maslow refers to these needs in his *hierarchy of needs*, often seen depicted as a pyramid, as shown here.

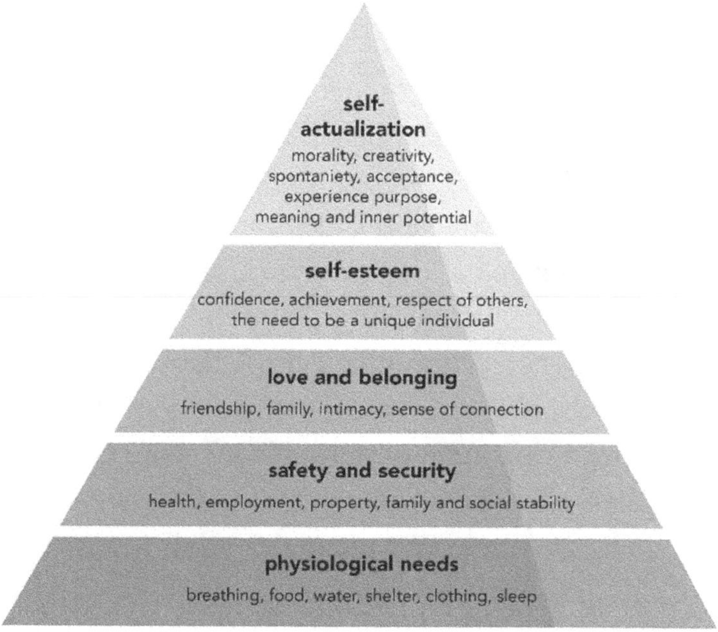

self-actualization
morality, creativity, spontaniety, acceptance, experience purpose, meaning and inner potential

self-esteem
confidence, achievement, respect of others, the need to be a unique individual

love and belonging
friendship, family, intimacy, sense of connection

safety and security
health, employment, property, family and social stability

physiological needs
breathing, food, water, shelter, clothing, sleep

Figure 6. Maslow's Hierarchy of Needs

Along the bottom of the pyramid are the most basic needs, often referred to as physiological needs. Those must be taken care of before we can seek higher needs, according to Maslow. After health, employment, family, and social stability have been attained, there is room for the need to feel confident, to achieve, to be respected, to be a unique individual, and even to experience purpose and inner potential. These latter experiences represent self-actualization.

Although only the first rung is labelled with reference to the physical body, each of the levels represents a physical component. Consider the experience of being ostracized at work, at school, or generally from your social group. At one point or another, many of us have gone through that devastating experience. I remember one year, in grade 5, when it happened to me quite strongly. I had just finished a rather successful year of being one of the most popular girls in grade 4. I was cute and friendly, a leader, and I was fun. Something changed suddenly in grade 5. Actually, I know very well what happened: one day I wore a fruit-patterned outfit including stirrup pants and a matching sweatshirt. I might have survived a social catastrophe had I worn either the pants or the shirt separately, but the combo killed me. I dropped so low on the social totem pole that one day in class I looked down and saw over a dozen giant spitballs all around me. I was beside myself with embarrassment, and I tried to ignore the fact that I was being attacked. I lost all my friends for grade 5 and was forced to spend a few lunch hours alone before I was finally accepted into a new, albeit unpopular, group of girls. My need for intimacy and friendship motivated me to befriend these girls even if it meant a further hit to my social status. Having friends and social connections was an important need, and still is for most of us social-by-nature humans.

When I think back to that time, I can still conjure up the body sensations, particularly when I think about the spitballs. Tears behind my eyes. Cheeks saggy with loneliness. Face blushing red. Heart sinking in. Fear of the ridicule I would face going out for recess. It was horrible in both my mind and my body. I could go on. The point is that our needs, even those non-life-threatening ones are real, sensational, and physiological. As such, they become powerful intrinsic motivators.

Self-Reflection Exercise: Needs Not Met

What does it feel like when your needs are not met?

What sorts of body sensations do you notice when you are feeling taken advantage of?

Notes:

Self-Reflection Exercise: Intrinsic Motivation

Assuming your basic needs were met as a child, you may have been able to seek other psychological goals. You were able to find activities that provided you with lots of reward, activities that you wanted to do for their own sake, activities in which you found sincere enjoyment. When you think back to your childhood, what do you remember about those activities? Describe those activities. Reminisce about those activities. Let your mind wander about those activities. Make sure to make a mental note of how you felt (including any physical sensations) after you let yourself indulge in those memories. Then ask yourself, how many of those activities do you continue to do today? If it's a lot, do you also feel satisfied and happy with life? If only a few, do you have the sense that doing more of those could bring about more happiness? Do you think you have an inner motivation to do those behaviours and that you are missing out by not doing them?

Notes:

I can also think about a time one of those needs was met for me. It started when I began studying neuroscience during my undergraduate university years. I remember sitting in the lab where I volunteered and talking with my professor about the brain. I had her all to myself, and I had a million and one questions to ask, one after the other like rapid fire. I was so enthralled with the answers that the questions just kept coming. I felt giddy inside. I felt teary-eyed, but teary-eyed like when I witness a beautiful wedding ceremony or see a puppy licking a kitten. I also felt a little bit of pain in my head, as if my brain were bursting out my body trying to get out all its questions about itself. My heart was racing. I was hot. I was aroused. My desire to know more burned inside of me. I was experiencing passion, and it was amazing.

I remember being acutely aware of this experience even at the time. I remember feeling that these physical sensations were so incredibly strong that I needed more of them. Neuroscience became my drug. I yearned for it and I loved it when I had it. These feelings motivated me in the pursuit of my goal. At a practical level my goal was to earn my PhD in neuroscience and psychology, but on an immediate-fulfillment level the goal was to learn, and that expressed itself through taking classes, volunteering in a research lab, and spending time with those who seemed to know, like my professors. This distinction is important to recognize because sometimes those seeking passionate careers spend too much time focusing on the final destination (like the PhD) rather than the next immediate step (learning more). Doing the PhD was so far away that if I hadn't enjoyed the journey (classes, lab work, etc.), then the path would have been long and arduous. But it wasn't. As cliché as it sounds, the journey is actually the reward.

These same feelings motivate each of us uniquely to pursue what feels right, purposeful, and meaningful. Take Navey Baker, for example. She spent much of her childhood acting like a dog. Her dad said that he would watch her run around the backyard on all fours for hours, seeming to believe that she was a dog. Her parents got annoyed eventually, but she never let it go. She loved her identity as such. Today as a 20-something, she no longer behaves like a dog, but instead, she is a tiger mascot, and one of the best around. Her story is zestfully passionate, and it brought tears to my eyes to hear about her. So empowering. So raw. So real. So determined to fulfill her needs.

Self-Reflection Exercise: Needs

Consider when your needs have been met. Can you identify the physical experiences associated with that need? Also, what were you motivated to do in order to meet that need?

Notes:

Animals, including humans, often work hard to achieve their goals especially when motivation is high. But there is another level to our goals in terms of what motivates us to achieve them. As Maslow described, once we achieve the basic physiological and safety needs, we seek to achieve other less tangible needs like love, acceptance, aesthetics, creativity, and even a need to express ourselves fully (the need to be self-actualized). These needs rely on others and are, consequently, affected by others. In fact, the external rewards and reinforcements we get from others and from our past also seriously influence our motivation. Some people might call these extrinsic needs because they come from outside, but I and others would argue that needs are needs, and that all have an internal component. From a neuroscience perspective, no matter what, they all end up emerging internally, through the brain.

The Neuroscience of Motivation

In animal labs we study motivation through a concept called goal-seeking behaviour. When there is a goal, say getting food or escaping, an animal can navigate through its environment and perform a variety of behaviours to achieve its goal. For example, a lab rat can be motivated to learn a complicated maze when it knows there is a tasty treat (the goal) waiting at the end. The prize motivates the animals to seek and achieve the goal. A lab rat can be motivated to learn the same complicated maze if, perhaps, it was put on a diet and it knows there will be an abundance of food at the end. Similarly, a lab rat might be motivated to find a hidden platform (the goal) submerged in a tub of water, in order to escape the chilly water (the motivation). The same rat might also be motivated by the safety of a dark, enclosed room (the goal) when it smells a predator, like a cat. In this situation the need for survival by escaping the predator becomes the motivation. A rat may also be motivated by the prospect of a sexually receptive female (the goal) with the scent of female rat pheromones (the motivator by virtue of implying sexual reproduction). Some animal behaviourists (and I would include myself here) would argue that animals are even motivated by a basic curiosity. Curiosity allows an animal to move beyond its safe environment and venture out into the scary world in order to find better shelter, food, water, or a mate to reproduce with.

Despite the wide variance in types of motivators and needs, animals are hard-wired in their nervous systems to seek goals through the "dopaminergic reward system" of the brain. This system involves a brain structure known as the "striatum," located in the middle of the brain along with another specific area of the striatum known as the "nucleus accumbens." The striatum is an area of the brain where the neurotransmitter dopamine is active. The striatum receives its dopamine from neurons that originate from a place farther back in the brain (known as the ventral tegmental area, or VTA).

To give an example of how powerful this system is, if we implanted stimulating electrodes into the nucleus accumbens of a rat and allowed it to stimulate that area of the brain on an unlimited basis, releasing dopamine, the rat would do so continuously, even to the point of killing itself. Stimulating this area of the brain is highly rewarding, and the VTA is the same area that drugs like cocaine and amphetamine work on.

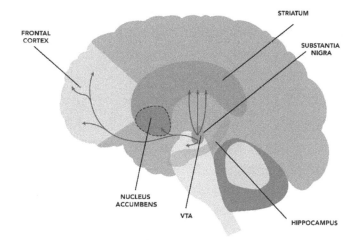

Figure 7. The two main dopamine pathways of the brain related to: 1) reward, including the origin of dopamine neurons (i.e., cells bodies) in the ventral tegmental area (VTA), which project axons up to the nucleus accumbens and 2) movement, including the origins of dopamine neurons in the substantia nigra that project up to the striatum. The hippocampus is shown as a landmark brain structure.

Dopamine is often referred to as the "reward hormone," but it would be better to think of it as the "wanting hormone." When dopamine and dopaminergic neurons are active, animals (including humans) feel a strong urge *for* something, that is, they "want." This is why the rat will keep stimulating itself. It wants more and more and is never fulfilled. It's similar to my wanting more and more answers as my neuroscience passion was ignited. However, this wanting is typically accompanied by the release of another class of neurotransmitters, known as opioids. Most of us have heard of endorphins, like the runner's high. Endorphins stimulate the opioid brain system. Drugs that work on those neurotransmitters include morphine, opium, and heroine. Opioids are "feel-good" chemicals and help give us feelings more like bliss. For me, when things made sense, when after all of my question-asking I felt I "knew" enough, I felt blissful and good. The reward associated with opioids, coupled with the desire associated with dopamine, becomes a very powerful motivator. It can even become addictive. Like I said, neuroscience and psychology became my "drug" because literally, neurochemically, I was feeling it. I am motivated to learn with dopamine, and I am rewarded with knowledge and opioids.

Self-Reflection Exercise

Spend five minutes jotting down answers to the following questions. Consider repeating the activity a few days in a row.

What do you really want? When do you want it? How does it feel in your mind, body, and brain?

What makes you feel blissful? How does it actually feel in your mind, body, and brain?

Do your answers to either of the above get you to move? Do they motivate you in any way?

Notes:

Shaping Our Behaviours, Beliefs, and Brains

If we were left to our devices (our internal motivation and drive) to secure our needs, we would probably do just fine. But our movement through life is highly influenced by our experiences and by the people around us. Those experiences and people shape our behaviours, our beliefs, and our goals.

Around 2006 I began blogging and posting other written commentaries online, on Facebook, and through other social media. What started as a few thoughtful sentences progressed into paragraphs and eventually into longer blog pieces and I soon began publishing articles in magazines. I feel a sincere reward from writing and am deeply and internally motivated to articulate myself through words, either spoken or written. I can spend hours or days or even years trying to perfectly articulate something. It is a dopamine-mediated behaviour of wanting. The gratification I feel when I think I have it nailed fills me with endorphins. Getting positive feedback on something I am already internally motivated to do is extremely powerful, and it's how I ended up typing these words. Each time someone comments on something I write, I am encouraged to write again. I am rewarded. My behaviour is reinforced, positively. This positive reinforcement (reward) increases the likelihood of my behaviour continuing into the future.

If, on the other hand, my work is met with harsh negativity or criticism, it decreases the likelihood that I will continue writing, not wanting to experience this kind of punishment. And that's exactly what happened for a while. In my fourth year of university, one of my profs sat me down in his office, handed me back one of my papers, and began to politely and sternly explain how horrible my writing was. I was awestruck. I actually had no idea it was that bad. I guess I had received false feedback (high grades in writing classes with no other feedback). I felt humiliated, like a complete failure. All the sensations of failure crept into my body, and it was terrible. In some ways, that experience prevented me from identifying as a writer for a very long time. But my relationship with failure allowed me to get back up. After that meeting, I went straight to the bookstore and purchased a grammar book, which I read immediately. Learning that I was not a skilled writer was great because it motivated me to improve. It wasn't the last time I was criticized, but it was easier each time.

Negative reinforcement is another way our behaviour is shaped by experience, but the concept is often misunderstood. Negative reinforcement is not just getting negative or constructive feedback. Negative reinforcement can also increase the likelihood of a behaviour happening in the future, just like positive reinforcement does, except with negative reinforcement, what happens is that something negative stops happening and that encourages us to continue our behaviour. I experienced a good example recently when I rented a car for the weekend. It was a FIAT and it made the most obnoxious sound when I began to drive without putting on my seat belt. I admit, I usually jump in a car and put on my seat belt as I start driving — at least until I rented a FIAT. The sound was loud and got louder the longer it was ignored. Over the course of the weekend, I learned to put on my seat belt promptly *before* beginning to drive away, just to avoid that awful sound. My behaviour was effectively shaped by negative reinforcement.

Another example of negative reinforcement, would be if, for example, as a writer, I suffered from headaches and felt unwell when I did not write. If beginning to write cleared away my physical discomfort, I would be more likely to continue writing in the future. This behaviour would be shaped by negative reinforcement. In that case the removal of the negative experience (feeling unwell) reinforced that I should keep writing to continue feeling well. Indeed, I can say for myself, it does happen to some extent. When I don't pursue at least some of the endeavours I am passionate about, I do feel unwell, uneasy, and generally out of sorts. I feel like I am missing a piece of myself and it doesn't feel good. Yoga, writing, time alone to reflect, and physical activity are some of my needs, and when I don't allow myself to do those, I don't feel good inside.

Self-Reflection Exercise

Consider the following situations and recall at least one experience of each.

> ➢ When something you did was strongly reinforced positively, increasing the likelihood of your behaviour occurring again in the future

> ➢ When you were punished for something you did, decreasing the likelihood of your behaviour occurring again in the future because of a negative consequence

> ➤ When something you did was negatively reinforced, increasing the likelihood of your behaviour occurring in the future because of the removal of something negative

Notes:

From a behaviourist perspective we consider this reinforcement and punishment in terms of the stimulus (S), response (R), outcome (O) relationship. When social media emerged (the stimulus) and gave me a new forum to write (my response), I was rewarded with the appreciation I felt from others (the outcome). This increased the likelihood that I would write again in the future and I did. Alternatively, I may have had the same stimuli (social media platform) and responded the same way (with writing), but the outcome could have been different. Instead of praise, I may have been criticized (outcome), and then my behaviour would have been less likely to occur in the future.

In the laboratory we examine this shaping using a paradigm known both as *operant conditioning* and as *instrumental conditioning*, denoted by the formula $S + R = O$ as referred to above. This conditioning is a type of learning, where the animal's own behaviour is instrumental (hence the name) in achieving the reward. What an animal does (its response) when offered a command (the stimulus) has an associated outcome (reward or punishment). This is the basis for positive reinforcement training with dogs. For example, a dog quickly learns that it will be rewarded if it responds to the owner's request to sit. The "sit" command is the stimulus, the action of the dog is the response, and the outcome is a reward if the dog responds correctly. The outcome is highly motivating for the dog because the dog's own behaviour is instrumental and, as for humans, can come in many forms: getting petted, being played with, being fed dinner, or, most frequently, receiving the dog treat.

Instrumental conditioning, as it is referred to in psychology and neuroscience, is different than classical conditioning. In classical conditioning, we have Pavlov's dog as an example, where the dog just sits there and its brain happens to detect an association between the bell and the upcoming meal. The dog and its behaviour do nothing to facilitate this. They simply coexist. Yet the dog's behaviour is shaped because it starts to salivate when it hears the bell, a behaviour that does not normally occur in nature. His behaviour has become conditional on previously learning that the bell and the food come together.

We can continue to let our behaviour, beliefs, and brains be shaped by our experiences, and inevitably this kind of response will happen. But we can also bring in some intentionality and consciousness to how we let those experiences shape us. For example, we can reward ourselves for going to the gym, writing, taking a course, or getting to work on time. If those are things we don't intrinsically want to do, then they become "shoulds" on our list — not always a bad thing if they drive us to be healthy and well. But we may need some rewards to encourage that behaviour. We can also focus on what internally motivates us and commit to doing more of those things in our lives, whether it is more creative work, more socializing, more learning, more sports involvement, more music, or anything else that turns you on, literally. It's not hard to do those things because we self-generate all of that internal goodness associated with dopamine (wanting) and opioids (feeling good). We just have to start and then let our bodies take over. We can also be more intentional about how those around us influence us, but that will require a bit more discussion.

Self-Reflection Exercise

How have others shaped your...

> behaviours?

> beliefs?

> expression of emotions?

> brain?

Notes:

You may come up with common examples if you think about peer pressure as a teenager, career choices, who you date, what you do or don't stand up for, where you live, what you buy, and what activities you engage in.

Shaped by Others

Some of our motivation comes from within, but some arises from external factors, particularly the people around us.

Parents

As incredible a job as parents do raising their children, they can often end up being huge obstacles to their children's happiness. One of my clients spent years in university avoiding art school because his parents told him that he could do whatever he wanted except it had to be financially secure, and it could not be art. Art school, from their experience, was not going to lead to a financial secure career. With good reason my client's parents believed this. His parents grew up in poor families, and one of their parents had been an artist, a real starving artist for a long time. They grew up associating the artist's way with poverty. As a result of a need for security, the parents drove themselves to success financially, but the outcome was a strong aversion to any of their children having to live in poverty. That subtle and not-so-subtle message was conveyed to my client, who spent a great deal of time circumventing a creative career. Two engineering degrees later, he finally went to art school and was in his glory!

There are many examples I could recount, but one other sticks in my mind. I had a client who was an engineer. She was depressed and very unhappy at work. We talked a lot about what she liked doing, trying to bring a little pleasure and passion into a life that felt meaningless to her. Often a good positive-psychology strategy is to focus on the good and not the bad, and to activate some wellness, but we couldn't come up with anything after thinking long and hard. She was apathetic about almost everything. I finally asked, "What did you like in university?" Without hesitation, she went on about her psychology courses and how much she loved them. I asked, "So why did you do an engineering degree?" Her response was common: "Because my Dad said I should." Sad. True. Common.

Teachers

We also hope our teachers encourage us to develop our talents and skills. Often this happens, but often it doesn't. I can't count the number of people I know who have had experiences of being told by teachers that they were stupid, incapable, wrong, or simply could never achieve what they wanted to. Those words inevitably lead a kid to avoid certain elements of themselves because they feel ashamed. A good friend of mine is a brilliant designer with exceptional spatial skills. His teachers told him that he should not go to university because he was not smart enough, a claim supported by his diagnosis with a learning disability.[47] For a long time he listened. He believed he was not capable and should not go. At the age of 27, he finally got the courage to apply, and he went to design school,

---

47. I prefer to use the term "learning difference."

where he thrived, graduated with a degree, and went on to start his own company.

It goes without saying that many others in our lives influence and shape our behaviours, decisions, beliefs, and even what we think we like. Partners, friends, co-workers and colleagues, teammates, mentors, children, and anyone else whose acceptance we want can ultimately influence us with the beliefs they hold and explicitly or implicitly impose upon us.

Media

Blaming the media is easy! A prime example is the incredible increase in eating disorders in young girls and women who falsely believe they are supposed to look a certain way — a way that is not even real, but a digitally altered fabrication. The media can influence our career aspirations too. As a kid I loved my sports — I might even say I was passionate about them — but I never considered a career in sports. Why not? Probably because there were few opportunities in sport for women and, the media made this quite clear as I was growing up in the 80s. Imagine not being able to see highlight reels for inspiration or be in awe of people with whom you could identify. There was no WNBA and women's hockey wasn't even in the Olympics yet. Imagine a time of incredible inspiration even for a non-athlete, but also imagine never having your sport represented, and never considering that it may one day be there. Imagine never seeing a female professional athlete, a coach, a sports reporter, a general manager, or a club owner in the public. Imagine being made fun of for the sport you played. Imagine not being welcomed on the ice at the community club. Imagine being told that "girls can't play, run, or throw." Imagine having no role models. That was my life as a girl who played sports.

Imagine the opposite, which is slowly taking form. As the first female skater in the European league, Hayley Wickenheiser showed us that women could play professional hockey.[48] Christine Sinclair, Team Canada's women's soccer captain who helped her team win a bronze medal at the 2012 London Olympics, has become an icon and a role model for girls everywhere, particularly in Canada. Becky Hammon became the first female full-time assistant coach in the NBA. Each of these women will be shaping the imaginations of young female athletes for years to come simply because they exist and serve as examples of the possibilities for girls.

Beyond sports we have great examples too. Barack Obama serves as a great role model for promoting success among black children. Hilary Clinton is already a role model, but imagine what her becoming the president would do for young girls' perceptions of opportunity. Actress Laverne Cox, who played a transgender person on the Netflix series, *Orange Is the New Black,* is transgender in real life and serves as a great role model in the media. In the summer of 2015, Caitlyn Jenner's *Vanity Fair* cover photo and story also made waves and, in so doing, makes her a role model for other transgender people and for the acceptance of that experience. There are many examples of how the media shapes our beliefs, behaviours, and brains in both positive and negative ways.

---

48. Manon Rhéaume was the first female to play in the NHL as a goaltender during an exhibition game with the Tampa Bay Lightning in 1992.

Being Aware of Our *Shoulds*

Our needs, goals, beliefs, opinions, and motivations can actually be quite complex if we try to break them down into our internal motivation, drive, and the ways others shape them. As we grow up we also learn what behaviours are more likely to earn us acceptance, companionship, social security, and so on. This inevitably gives rise to a list of *shoulds* that can direct (or motivate) much of our behaviour. These *shoulds* include behaviours, thoughts, and emotions that we feel we *should* engage in. For example, we might think we should get a job that pays lots of money. Or that we should become doctors. Or even that we should go to university, get married, have children, buy a house, love someone of the opposite sex... By the time we are 30, we might have lengthy lists of things we believe we should do, think, feel, say, or experience. These *shoulds* come from a message from outside, not a message we were born with. They correspond to the expectations others place upon us, coming largely from several highly influential entities.

These external factors often pop up as a lengthy list of *shoulds* that entertain our minds for a long time. I should go to the gym. I should eat well. I should send thank-you cards. I should work hard. I should go out and be social. I should be the team lead. I should have a marketing plan. I should call my friends and family more. I should save more money. There are many, many *shoulds* on our list. What's on your *should* list?

Self-Reflection Exercise

Write out your *should* list.

Consider where the source of each *should* comes from. Is each one internally driven and does it represent a true desire on your part? Or is it externally motivated and shaped by others?

Take a moment to consider what factors outside of yourself motivate you to do the things that you do in life, in your day, in your relationships, and at your job.

Notes:

External factors are often powerful motivators and can both impede our ability to achieve and propel us toward something inherently important to us. Being surrounded by people who want a particular fate for us (for example, to maintain a certain lifestyle, to go to university, to work in a particular field, to date a certain type of person, etc.) creates an inner struggle and can drain our energy, particularly when that fate does not coincide with what we want for ourselves. Imagine a huge amount of resistance when we attempt to pull away from these expectations and seek out something more inherently of interest. It can be hard to separate the rewards and needs of social acceptance from internal self-actualization needs. Alternatively, being supported and having internal and external needs aligned can be powerful in terms of helping us achieving very meaningful goals. Compare situations in which parents are or are not incredibly supportive of their daughter's artistic career choice. Or compare scenarios in which parents are incredibly supportive of their son's homosexuality in one case and parents are not supportive in another. Or, imagine a situation in which a husband is incredibly supportive of his wife's goal to climb a corporate ladder compared to one in which the husband is not. In each of these scenarios we can imagine how difficult and energy-draining the unsupportive situation is, all because of the motivation to feel accepted and loved by those around us. Consequently, taking stock of our social environments is often helpful in accessing our ability to succeed. And if, during the process of taking stock, we realize that our social environment is more draining than nourishing, efforts could be made to change up our social environment by joining communities of people with similar goals, finding a mentor, or helping others see what things are important to us. It doesn't mean we have to get rid of people who influence us in a way we don't want, but rather that we can become aware of the impact and strive toward bringing in more positive influencers to support the direction we do want to go in.

Curiosity

One very strong motivator of animals, including humans, is curiosity, otherwise known as "novelty-seeking." We all are innately driven to seek novelty. In fact, we use a variety of mazes in the laboratory that rely on an animal's innate curiosity, curiosity that motivates the animal to explore and move through its environment, particularly when the environment is new. Without that curiosity it would actually be hard to get an animal to move, and without movement we could not test things like memory, motor skills, sensations, perceptions, fear, or motivation, so we would not be able to study disorders like Alzheimer's disease, amnesia, stroke, Parkinson's disease, anxiety, and depression, or skills like vision, smell, and hearing. Each of these disorders and skills is assessed by observing basic animal behaviour, or the basic animal curiosity for exploration.

One test used in the lab is called the elevated plus maze. Just as the name indicates, the maze stands elevated above the ground and includes four planks extending from a centre space to resemble a plus sign. Two of the arms are enclosed with high arms, forming dark alleys, and the other two are open, high above the ground. To start the test, a rat or mouse is placed in the centre of the plus, and the experimenters watch to see where the animal goes. A typical rodent will venture quickly into the closed arms to feel safe and secure, demonstrating normal, baseline fear behaviour. Rodents don't like to be exposed in the open because it's not very adaptive in terms of protecting oneself, for example, from vultures from above. But a typical rodent won't stay in the closed arm for long. Soon the fear dwindles, and the rodent is motivated more strongly by its own curiosity to explore the rest of its environment, the maze. After having explored the first arm, a typical rat will make its way back to the open centre, peek around the corners to see the open space below, then quickly run across to the other closed arm, and begin its exploration of that relatively safe arm. I use the word "relatively" because there is some inherent fear present as the rat moves about, but as it explores, the rat becomes more familiar with its environment and the fear dwindles further.

Humans are similar. Recall the last time you were in a new environment. Perhaps you were at a party with new people. Perhaps you were starting a new job in a new office building. Perhaps you were visiting a new city. Each of these new environments comes with a bit of fear or anxiety. That's natural. But over time, as we become more familiar with the environment, under normal circumstances we settle in and feel more at ease. The same response occurs inside the brains and bodies of rats exploring the plus maze.

Eventually, animals become so comfortable in a new environment that they will keep venturing out. In the plus maze, the animals head back to the centre, peek around the corner again, and look out to the open arms of the plus. The open arms are the scariest of the entire maze, at least at the beginning. They are high and exposed, making the animal easy prey out there. But curiosity typically gets the best of them, and they eventually make their ways out onto the exposed open arm. Often a rat will stand in one spot along the arm and dip its head down to see what's below and to assess how far it is from the ground. Through this observation, the scientist is recording many details to capture the curiosity: 1) length of time until venturing onto the open arms; 2) number of peeks around the corner and for how long; 3) number of head dips looking down off the ledge and for how long; and 4) number of times moving all the way to the end of the arm. Collecting these data forms the basis of comparing curiosity — or rather the degree to which curiosity emerged and fear subsided — among groups of animals. In my own research, I did this for the purpose of testing antianxiety treatments, including both pharmaceutical medication and herbs. When a treatment worked and effectively reduced anxiety, the rats would more quickly venture out into the open arms, peek around the corners, sit on the open arms and dip their heads, and move all the way to the end of the open arms.

This curiosity-fear continuum is so fundamental to animal behaviour that it is used in many behavioural tests in psychology and neuroscience laboratories. For example, a maze with food at the end requires the animal to be curious enough to move from the starting spot, walk into a blind alley, walk back out, and then attempt new alleys, eventually happening upon the yummy treat at the end. Several days may go by before the rat can easily run right from the starting location to the treat. This is how we test memory. Curiosity gets the animal motivated to explore the environment in the first place. That exploration leads the animal to learn and remember its environment and continue to move so it can get the yummy treat at the end of the maze. But, if an animal has memory problems, it won't be able to remember where to go and instead, might take many wrong turns along the way.

Once an animal has been motivated by curiosity to move and explore and learn about its environment, we can test all sorts of features of memory. Does a particular drug improve memory? Does brain damage in a certain region impair memory? Does another particular drug destroy memory? Are males better or worse than females? Are older animals better or worse than younger animals? Does having lots of siblings or growing up in a complex and enriched environment improve memory? Does having a good mother improve memory? The questions are infinite, and they all rest on the ability of the animals to be motivated by the very basic drive of curiosity.

Curiosity as an Opportunity for Personal Growth

Todd Kashdan, Paul Rose, and Frank Fincham,[49] researchers in the Department of Psychology at the State University of New York, indicate that curiosity is an opportunity for personal growth. How can it be?

Self-Reflection Exercise

Consider the "elevated plus maze" as an analogy to what we avoid and approach.

Closed:

What exists in your dark alleys? Where is your safety zone?

Where do you run and hide when you don't feel safe? This might consist of people, places, routines, familiar experiences, work, etc.

---

49. Todd B. Kashdan, Paul Rose, and Frank D. Fincham, "Curiosity and Exploration: Facilitating Positive Subjective Experiences and Personal Growth Opportunities," *Journal of Personality Assessment* 82, no. 3, (2004): 291–305.

Open:

What exists out on your open arms?

What opportunities are there for you to explore that feel just a bit uncomfortable but are intriguing enough that you go there anywhere?

What are you curious about and motivated to seek that will ultimately allow you to expand and to give rise to personal growth? This might consist of any possibilities that are within your reach, but lying outside your comfort zone.

Centred:

What is your centre? What is the place where you can equally move between safety and exploration? Where do you feel in control of your destiny? This might consist of people you rely on, routines that make you feel secure, and any other safety measures that still allow you to feel compelled to look out onto those open arms.

Notes:

Curiosity and the Brain

Most of the neuroscience research on curiosity is studied under terms like "novelty seeking" or "exploratory behaviour." In each of these cases the difference in the brain between new experiences and old and familiar ones is fairly well understood. Activity within the hippocampus, frontal cortex and, more recently, structures of the dopamine-reward system is typically the root of this difference. The hippocampus and its surrounding structures are highly involved in our ability to create memories, in particular, of our environment. As we walk through an environment we are making use of the hippocampal mapping system, which includes cells specifically designed to code where we are in space (called place cells and boundary cells). These cells are activated and form a gridlike map for us as we walk through an environment. If the environment is new, these cells begin to code the features of the environment and retain the information in our spatial memory. If the environment is familiar, cells that previously coded the features are activated and give us that sense of familiarity. For lab animals, this happens when placed in environments like mazes.

The places cells help an animal move from a state of fear and anxiety to a sense of familiarity. This same thing happens to humans as we enter a new environment, for example, moving to a new city or starting at a new school or workplace. At the beginning, the exploration is stressful and provokes anxiety. But as our brains adapt through a familiarization process, the situation becomes less stressful and the desire to be curious grows again. There is a constant cycle between curious behaviour and familiarization. We never want to be in a state of too much or too little novelty stimulation. Ideally, we hit a balance and are able to securely and confidently explore life.

Another important brain element of curiosity has to do with dopamine, the neurotransmitter involved in wanting and reward. Remember, dopamine is the neurochemical that is targeted by very powerful stimulant drugs like cocaine and amphetamine, and without adequate dopamine, a person (or animal) can suffer from a lack of motivation. This connection between dopamine and its associated brain structures (for example, caudate and nucleus accumbens) was described in a study by Kang and colleagues,[50] who found that motivation to learn trivia answers was associated with greater activity in these regions of the brain. In a 2014 study, researchers Gruber, Gelman, and Ranganath[51] discovered two important findings related to curiosity and learning. The first finding was that states of high curiosity enhanced learning but not just learning of the information that was interesting to the participants of the study but also of incidental information. This is important when considering the value of curiosity, particularly whenever we find ourselves in a teaching role, such as teaching school, parenting, or training someone in the workplace. If we can invoke in our students a sense of curiosity, they are

---

50. Min Jeong Kang et al., "The Wick in the Candle Of Learning: Epistemic Curiosity Activates Reward Circuitry and Enhances Memory. *Psychological Science* 20, no. 8 (2009): 963–973.

51. Matthias J. Gruber, Bernard D. Gelman, and Charan Ranganath, "States of Curiosity Modulate Hippocampus-Dependent Learning via the Dopaminergic Circuit," *Neuron* 84, (2014): 486–496.

likely to learn better even if they are not particularly interested in the material!

The second finding of that study emerged when the researchers looked at the brains of the individuals in the study. Through fMRI (functional magnetic resonance imaging) they found that learning and curiosity were greater when there was corresponding activity of dopaminergic brain circuits, including the hippocampus. This finding is correlational, meaning we cannot infer one causes the other, so it is impossible to determine whether heightened dopamine activity leads to better learning and curiosity or whether better learning and curiosity leads to greater dopamine activity.

These results suggest that curiosity enhances learning and memory and that it does so through activation of specific dopamine pathways and through the inherent reward inspired by curiosity. For me, this also further supports the idea that curiosity is valuable enough that the brain devotes resources to it, and that curiosity has some adaptive value even if only to help us learn and remember.

The last major brain area involved in curiosity is the frontal cortex. The frontal cortex is activated when something new is being experienced. For example, during the early phase of learning a new task, there is a lot of activity in the frontal cortex, but once the task becomes familiar and it loses its novelty appeal, activity in the frontal cortex subsides.

Curiosity in Practice

Curiosity is fundamental to all of us animals and so it should be valued. The brain has devoted significant resources to novelty and exploration, so they must be necessary elements of our existence. But curiosity doesn't come easily to everyone. Curiosity is impeded by fear and anxiety, and by too much stimulation and novelty. In fact, fear, too much novelty, and too much stimulation can be so strong that they essentially paralyze us from venturing out. This is important to take into consideration when considering our own curiosity and then when planning our own curiosity-inspired adventures, be they travel, education, or any new activity. In order to promote curiosity, we need to feel safe in our environment, which can come in the form of routine and familiarity.

It might be useful to determine if our brains are prepped for engaging in more curiosity in our lives. Do we have the brain energy to do so? When I moved to St. Catharines, my brain was not in a state of readiness. I was too overwhelmed with novelty and had little capacity for any other newness, including any curiosity. But at other times, when we are in a state of the mundane and too much routine and we need a little fresh perspective to turbocharge us, we have a much greater capacity for curiosity. Taking that inventory is helpful before determining a plan.

Coming to value my own curiosity has been both a blessing and a curse. On the one hand, I am incredibly fortunate to have found many homes for my curiosity, including the laboratory where I spent many years attempting to understand a variety of different psychological and neuroscience phenomena. I have also found a home for my curiosity in teaching because just as much as I teach what I know, I also attempt to learn more in pursuit of remaining relevant. I also have a home for my curiosity in everyday interactions with people, with my clients, and with all other humans around me, as I am enticed to learn more about and from everyone. Each and every one of us is a goldmine to be explored! When I remember that, my social interactions become much more pleasant. I remember one time in particular when I found myself sitting next to someone who, initially, I could not relate to at all. On the surface I didn't think we would have much in common. But as time went on I recognized an opportunity to be curious. Indeed, within a few minutes of questions, I had tapped into a common ground and was fully engaged. I learned that he was incredibly passionate about his life on Vancouver Island and all of the activities that he takes part in. I learned that he takes great pride in smoking fish. His enthusiasm was contagious and eventually I found myself absolutely captivated with learning about how smoking could be done both cold and hot. Who knew cold smoking was a thing? Not I, and nor would I, had I not adopted a perspective of curiosity in this casual conversation.

Curiosity Self-Reflection Questions

Where is curiosity a part of your life?

What are you curious about on a daily basis? Do you read the news? Do you ask people about their days in a sincere way? What do you learn?

How does it make you feel to be curious in this manner?

What are you curious about on a grander scale? Did you enjoy your education? Do you travel? Do you wonder about the meaning of life?

How does it make you feel to be curious in this manner?

What if you lost your curiosity? Who would you be? How would you feel?

What if you had space to be more curious? What would you do?

Notes:

Valuing curiosity has many personal benefits, but by activating our own inner curiosity, we stand to be a bit more tolerant of each other and how others might do something differently from us, by simply asking "why would someone do that" in the most non-judgmental way possible. The element of curiosity is much more than a motivational drive to explore. It is a stepping stone along the path of compassion, self-awareness, and ultimately, self-actualization. It's always a driver in finding our passion.

Finding Your Passion

Where does passion fit amid all of this? Is it a need? Is it a motivator? Is there social pressure? As far as I have read, Maslow doesn't talk directly about passion as part of his hierarchy of needs, yet I do think passion is embedded within it if we adopt a broader definition. When we consider the top of Maslow's pyramid, we see a need to self-actualize, which implies a need to express ourselves as truly and authentically as possible, to literally become our true selves. In some ways, this need to express our uniqueness seems to contradict our need to assimilate and be accepted and loved by those around us. But remember the pyramid of needs is a pyramid, and according to Maslow we can only progress upward when we feel secure on the rung below. Therefore, feeling accomplished in the departments of social acceptance and love means we can confidently strive to achieve our need to self-actualize, as contradictory as that may seem at first. This is how I think Maslow intended his hierarchy of needs to be interpreted, and that passion is possibly the result of feeling aligned with our needs and being within the process of self-actualization.

In pursuit of self-actualization, we begin to pay attention to, and actualize, those idiosyncrasies that make us different. Those unique traits might come with a deep sense of gratification as we pursue self-actualization, a gratification that might feel a lot like what we think of as "passionate." When I think about my unique characteristics, I think about my love for women's equality, for playing sports, for writing, for educating, and for bringing about new states of awareness. I might also think about my love for animals, my need for alone time to reflect, and my deep, long-standing interest in the mind and how it works. Each of these can invoke in me a sense of what I describe as passion. The passion I feel from each of these is not necessarily what other people feel. My partner, Mike, for example, gets passionate about some of those things, which makes us quite compatible, but not all them, which makes him unique. My best friend Lindsey also gets passionate about some of those things, but she's also very passionate about many different things. Among my parents, brother, friends, clients, students, we all have unique interests, but we share some of them.

As we continue to strive for self-actualization, we learn to really honour these unique areas, and if we end up pursuing them with some frequency, we can feel a deep sense of reward and fulfillment, or passion. That is the nature of a passionate pursuit as I define it. It's nothing complicated. It's simply exploring who we are, uniquely expressed as the individuals we truly are. If we were to ignore these unique interests, these passions, and the intense sensations that turn us on, we would be ignoring our need to self-actualize. We would be ignoring big parts of who we are. In the process, we fail to achieve our ultimate need, according to Maslow, to express who we are most authentically. What a shame that would be!

As children (and as adults), we are constrained by the perceptions around us. Most of us would not say we wanted to be, for example, a mascot when we grew up. Navey Baker didn't either. When we are asked as children what we want to be when we grow up, we can't actually know all the possibilities — in fact some have yet to exist! — so we come up with the best approximation of that as a guiding light. We rely on our experiences and best guesses. This is why girls would say things like "teacher," "nurse," and "ballerina" instead of "football player," "cowboy," or "businessman." Girls don't see many cowgirls or female football players, but they do see female nurses and teachers and dancers. This is also why boys would say things like "hockey player," "pilot," and "firefighter" instead of "dental hygienist," "Avon representative," or "ringette player." On a side note, I do know of one little boy who said that he would *not* play hockey because "hockey was for girls." This was based on his experience of going with his athlete mom to *her* hockey games and having only she as the prototypical hockey player in his mind. Whether boy or girl, very few of us said that we wanted to be a "software engineer," a "border patrol officer," or a "policy analyst" if we had no experience of what those jobs are.

Our brains, as glorious, capable, and miraculous as they are, still cannot predict the future or fathom all possibilities. We must, therefore, rely on the accumulation of experiences to both get ahead and learn how to navigate this life. The passion we may be following, or the "true self," may in fact be right under our noses, but it may be hidden by our perceptions and perspectives. If so, the only way to unveil it is to change those perspectives. How best to change perspective? We must literally move our bodies so we can see from a different angle.

Becoming a dog may have been one manifestation of Navey Baker's passion, which served her well as a child. But she eventually found a place as a mascot. Is that her final destination? Maybe not, but who knows where the drive inside her will lead next. I suspect that it will be somewhere satisfying and joyful.

For me, finding my passion came easily, in some respects. I didn't flounder through years of university tuition and end up not knowing what I wanted to do. I knew, without a doubt, that I wanted to study psychology and then neuroscience. I felt it deep within my body. With intense sensations of yearning, I would well up with tears while learning about psychology. I could feel excitement behind my eyes and in my head. My heart would race. I would get giddy. It was like I was in love. Actually I was in love! As a child, I used to get the same passionate feeling when learning about things that were mysterious, like paranormal activities (for example, ghosts, dreams, and clairvoyance), my preliminary psychology. I had a deep interest in the mysteries of the world and the mysteries of the mind, but none of this was nurtured in school before university. As a result, I failed to thrive until I got to university where I could fully immerse in a whole new world opened up for me. I found a place where I could explore my passion.

Nowadays many people say that it is terrible advice to suggest that someone follow their passion. For example, Cal Newport[52] is a GenY'r who criticized the slogan (as he called it), "follow your passion." Newport says that those of Generation Y who are confused about how to follow their passion would be better served with information and "concrete, evidence-based observations about how people really end up loving what they do." Fair enough. But it's not just a slogan. Finding passion leads to a true sense of happiness that comes from expressing ourselves fully and from the journey toward that expression, both of which require finding what makes us feel alive in this world. Passion, that intense desire or enthusiasm for something, can be a very strong motivator. Learning what we are deeply passionate about can propel us through many hardships and give us energy we never thought we had. That makes life a lot easier and rewarding to experience. The challenge is not in having or getting a passion. It's in finding the confidence to actually pursue it. We all know it's there. I am 100 per cent convinced of that. But we need to learn how to remove the obstacles and grab it.

If I knew of a surefire, quick way to do this, I would be a millionaire! But I don't. Sorry. The simplest suggestion I can offer is this: pursue what you love by first doing the most obvious and easiest thing aligned with it. Several months ago, I was talking with a friend of mine who was in a rut. She was in a dead-end job, in conflict with her boss, and she was miserable. When I

---

52. Cal Newport, "Solving Gen Y's Passion Problem," *Harvard Business Review,* September 18, 2012, http://blogs.hbr.org/cs/2012/09/solving_gen_ys_passion_problem.html

asked her what she would do if she could do anything, she said she would write. My response? "So write!" To me, it's a win-win situation. We do what we love whether we get paid or not. And in the end, the activity serves us because it brings us joy. The journey really does become the reward, and if other rewards happen upon us in the process, it's all bonus.

This book and all the writing I do is also a great personal example of finding passion and enjoying the journey. My goal was to write this because I know I enjoy the process of writing and attempting to articulate myself. Ideally, the product reaps some benefits too, but even if not, I get the opportunity to work with ideas and learn more in the process. Again, it's a win-win situation for me. Sure, I hope to be successful with all my work, but if I don't, at least I can look back at my life and know that I enjoyed my day-to-day living. To me, that is less of a risk and a much more rewarding existence than striving for success while miserable. In fact, I have been that miserable person in the past when I stopped liking the journey I was on. That's when I knew I had to get out, and that's ultimately why I left my post-doctoral research position.

Lots of people don't do what I do. Others work to play, whereas I play at work. Neither is right or wrong, but it is worth choosing consciously. It's also worth enjoying some aspects of the journey, if not all. Several months after that initial conversation with my friend who wanted to write, I ran into her and asked how things were going. A smile grew across her face. She told me that she took our last conversation to heart, quit her miserable job, started a new part-time job that covered the bills, and began writing. As she spoke, she lit up more and more. She had found the time and energy she needed to do the thing she loved. Then she said to me, "You know, I love it. I make myself laugh daily with my own words. And, even if nothing comes of this, I have enjoyed the journey." My heart almost burst out of my body. I was so happy for her. I also have no doubt that she will be successful with her writing because she is talented, and I fully believe that when we immerse ourselves in our talents, we prosper. So the surefire way of finding passion is just to do what we love. It's incredibly simple, yet it continues to pose a challenge for many people who, I think, are overcomplicating the experience. Therefore, I offer these points of consideration to help ease the process.

Find Passion in Menial Activities. My own career path found me in many jobs along the way that may not have seemed, on the surface, like jobs I would be passionate about. For example, while going to school, I worked part-time in retail customer service. Although I didn't end up in a career in customer service, I could have, quite easily. I love customer service! To me, a psychologist at heart, retail and customer-service industries are very practical ways to study human behaviour! Now I continue to study psychology, but it has expanded into the study of mind philosophies like yoga and Buddhism. There are many ways I can interact with this passion, and this is not the only passionate pursuit inside me. In fact, there is a part of me that has always regretted not working as a server in the restaurant industry. I suspect that serving would offer yet another opportunity for me to follow my passionate desire to understand the mind at work.

Consider Your Needs. The biggest barrier to finding passion is too much emphasis on other people's expectations and needs. Essentially, this is like being stuck on the rung of acceptance, social security, and companionship on Maslow's hierarchy of needs. We have to remember that finding our passion has nothing to little to do with our "should" list.

Look for a Mentor. Strutting down our own paths is difficult enough as it is, but without having seen anyone do it (or something similar) before us, the journey is that much more difficult. Mentors help us feel secure and not alone. They give advice based on their experiences. They reduce some of the effort on our part so we can move with less resistance. Finding one is like finding a gold mine and not having one feels like having a huge marble slab in front of our path. Having mentors also helps us achieve that need for social acceptance. Similarly, it's helpful to recognize how mentors we do have might turn us away from our passions, like parents or partners with their own agendas who want something very different for us. That's an example of an opposing mentor. We need to balance out the promoting mentors with the opposing mentors.

Secure Baseline Finances. The reality is that finances can stand in the way of our passions. But sometimes the perception of financial constraints is more of an obstacle than the finances actually are. One of my clients was working at a desk job she hated. She wanted to be outside and doing something meaningful, but she believed that it wasn't practical with all her student debt. With some work, we first got her on a self-care plan, which included regular relaxing baths. Although she felt guilty about taking the baths (for environmental reasons), the relaxation permitted her to indulge herself and cleared away enough resistance that she could clarify her next step. As a result, she ended up applying for a job with the Vancouver Olympics. That all happened in three sessions, and then I stopped hearing from her. A few months later I got an exciting postcard from Vancouver saying that she was working with the Olympics and was so thankful she had found her way!

Value What Comes Easily and Naturally, Especially If It Scares You. A lot of times people fail to see what is right in front of them because it is so natural and obvious that they don't realize the uniqueness of it. I was working with one client who wanted to explore more art but didn't value it as something special. She figured everyone wanted to do art. I told her frequently, "No, not everyone wants to do art. I don't!" which is the truth. I have no interest in art, so much so that when I go into art or design stores with my partner, Mike, I am bored. Though I try hard to be there supporting him, I get nothing out if it myself. Nothing, not one ounce of passion is evoked inside of me. It is nonexistent. It's hard but necessary to convey this to other people who often find themselves in communities of like-minded people. To them, everyone seems to like what they like, when in fact, not everyone does. While working with this client on pursuing art, she ended up finishing a design piece she had started a while back and she posted it online. She told me that it really scared her to do it. I believe that when things scare us in that way, it's because we are presenting a piece of ourselves to the world, a part of our natural and true selves. I told her about the things that scare me, like posting my writing. It scares me because it is dear to me and to my heart, just as posting her design work is dear to her and to her heart. Certain things pull at our heartstrings, while others do not. What pulls at one person's heartstrings doesn't touch another's. As a life coach, a question I might ask of someone trying to find their passion is "what scares you the most?" Then I would suggest going in that direction.

Sadly, many believe that their passions and dreams are a dime a dozen. They believe they are not unique in their interests. They believe there are many people out there with similar dreams and aspirations, and they don't bother investing in their interests. "Doesn't everyone want to travel?" a friend says to me. "Doesn't everyone want to be an organic farmer these days?" Indeed, no. I know many people who don't want to do any of those things. These desires are unique. Too often we take our dreams, passions, aspirations, and visions for granted. We assume because other people have them that they are less valuable. Not true. Those desires inside of us that wrench our hearts, burn inside, and fill us with an overwhelming sense of joy are worth more gold than the world has to offer. They give us energy and within them lies our true callings, the path to a meaningful life.

So ... value your passions. Don't discount them. Don't discard them and don't falsely assume "everyone wants to do that." If you want proof, go out and find 20 people who do not share your vision, and soon you will begin to appreciate the uniqueness of your own interests. And yes! some people do make careers out of travelling. Some people do become professional athletes, musicians, artists, and dancers. There are people who do garden for a living! Some people do spend most of their time outdoors. Or open wellness centres. And people do become explorers of the mind. Many people do. There are seven billion of us living in this world, so there's bound to be some overlap. But no two people will do it the same way and no one else is on the very same path as you.

Follow Your Paths, Not Someone Else's. Sometimes when we are confused about where to go to find our purpose, path, or passion, we latch on to others'. This might be one reason that inspirational talks like TED talks are so popular. A culture yearning for meaning, inspiration, and motivation sees inspiring people doing and sharing inspiring things. We get our shot of inspiration, but sometimes we are left stagnant. Where's the motivation? In fact, it is there. Inside us. It would not get activated if we didn't have an element of it inside. That's what feels so good. Those motivational speakers may activate our dopamine and our endorphins, but they don't always leave us knowing that those chemicals are inside of us — they aren't attached to someone else. It's like when we confuse someone else making us happy with someone else sparking our own inner happiness. All of these emotions exist inside of us.

Make a Decision. Indecision is a decision. Inaction is also a decision. So we need to be conscious of which of our inactions are making decisions for us and then decide whether that's the decision we want to make. If we make a seemingly "wrong" decision — and I personally don't believe these exist — we can learn to acknowledge it and move anyway in a new direction, having been informed by that "mistake." We have to recognize that our experiences and knowledge at the time of a decision were limited by that time, and we acted with all the information we had. Often, we are confused and stuck because we don't know where to go, and we don't know which option to pick. That sometimes leads us to do nothing in fear of making the wrong decision. Rather than becoming complacent, I suggest we pick a direction consciously rather than let one pick us, recognizing that nothing is permanent, and allow ourselves the opportunity to learn something new about ourselves, our interests, and our passions.

Learn to Trust Yourself. This is a big one to tackle and it's really the same as self-actualization. But consider it a process rather than a state of being. As such, learning to trust ourselves means that we take small steps in the right direction, get up if we fail (see the later chapter on failure), keep moving forward, and assess how well we listen to ourselves. Eventually, we will gain trust in ourselves. We have to. Because if we don't trust ourselves, who can we trust? I have made some pretty radical choices in my past; for example, deciding to go to Dalhousie for my PhD without even going to visit, leaving my laboratory science career to start my own business, moving to Toronto, starting a life-coaching certification course. Those were some big moves that I was able to make, largely on whims, because I knew myself and had built up a significant amount of trust in my process. I would not necessarily recommend choices like that early on for people who do not yet trust themselves. Rather, it's better to start with little intuitive guesses about what we should do (choices with little impact at first) and then see what happens when we follow them. If we guess right, we build up understanding of our own intuitive judgment and, in the process, begin to nurture a trust in ourselves. Eventually, we can make seemingly big decisions on a whim. More on that in the chapter on Intuition.

Flirt with Ideas. There are many things that we can do in this life, many ways in which our passions can manifest. Our job is not to do them all; our job is to recognize that energy inside of us and take action that aligns with it. I feel particularly blessed with this awareness and with the ability to pursue things I love. But I also feel blessed that I know that I cannot do everything. I have to choose. I often find myself in inspiring conversations with intelligent and motivated people where there is an instant chemistry of ideas, something that I am quite passionate about. In one week alone I might have a dozen or so such conversations. The other day I got an email from someone I had been conversing with. I had said something like, "we should write a paper on this." Indeed, it would have been fun, and I got caught up in the moment. But when he contacted me, I realized that I couldn't possibly follow up with this idea, as passionate as I may have been about it. The conversation was a mere flirtation with ideas rather than a direction I could pursue at that time. The truth is that there isn't time to do all the things I would love to do in this life. So I must admit when I simply *cannot* make time. Sometimes, I feel plagued by too much passion. We can flirt with ideas, dabble with what we like, but ultimately we must pick a direction once we feel ready.

Remember that Following Your Passion Is Easy. To follow our passion is actually quite easy. We just need to engage in (or become aware of) activities that we love and that make us feel good, activities where we lose ourselves and those that feel effortless. We should not delude ourselves into expecting passion to be present 100 per cent of the time, but rather, just enough for us to feel inspired in life to keep working at this greater purpose. And this, I think, is how people really end up loving what they do.

Passion Is Everywhere. I experience my passion often. The other day I heard actress Emma Watson address the UN about a feminism campaign aimed at reminding the world that feminism is about equality for both men and women. It made me cry several times. That's one of my passions, and those same physical sensations motivate me to fight for the equality of opportunity for both sexes in sport, science, and business. When I read an article on Oprah's school in South Africa, I was overwhelmed with tears of emotion. I felt so inspired by these young African girls who got a shot at an educated life. The article led me to think about my own work on educational innovation and reform. I daydreamed about my own university. I have been thinking about starting my own university since 1994, my second year of university. The idea manifested in small forms for the past two decades, and then I finally decided that I had to move on it. I realized that if I didn't, and if that opportunity passed me by, I would be very upset. This was a passion I had to follow and I'm glad I did.[53] It's the same with writing this book. Leading up to the last push to get this manuscript done, I was in an all-

---

53. That "university" only lasted two years and then we abandoned it, largely because it was not going in the direction that I liked. Sure, it was a failure, but definitely a good one because it taught me lots and I have no doubts that something similar will emerge over the next decade as a result of that trial and error.

time depth of depression. I feel it was my heart screaming at me not to give up on my dream.

There are many things that we can do in this life, many ways in which our passions can manifest, many forms of intrinsic motivation. Many young people nowadays struggle with this concept and much has been discussed on the topic. The best advice I have heard was by *Wired*'s founding executive editor, Kevin Kelly, while being interviewed on the *Tim Ferris Show*. Kevin Kelly said that young people should go off and do things without purpose and even more importantly, they should not be productive. Later in life is a time to be productive, but before that we are trying to learn lots and explore. We are trying to enjoy the journey and be in process.

Eventually, if we give ourselves the freedom at some point, we will settle on a few things that give us meaning and purpose. Once we recognize the energy inside of us, we can take action that aligns with it — the easiest next step. I feel particularly blessed with this awareness and with the ability to pursue things I love. But I also feel blessed to recognize that I cannot do everything. I have to choose. This is why we prioritize and set goals.

Passion Caveat

Passion alone is not enough for success. In addition to finding and following a passion, we must learn how to successfully accomplish our goals and achieve our desired passion-based career or life. Once we have taken the first step to understand what we are seeking then we can plan some actions in order to move in that direction. Goal setting is discussed in the next section of this book because too often we set goals that lack meaning, purpose, and passion, and we must understand our minds and brains better before we can legitimately set goals. I hope what we have covered so far has given us a good basis to move on.

Section III

Super Powers

I want to introduce the idea of super powers — super brain powers, that is. Yes, they exist. We have already covered some of what we know about how the brain works. By now we should believe in the brain's incredible power. But the brain and the mind are more powerful than we can appreciate. Although I don't think we know how to fully harness that power, I think we know how to capitalize on at least some of it. Up until now we have focused on some very practical super powers. We covered stress and relaxation as one super power along with the phenomenon of "cognitive load" in the chapter on stress. Reducing stress to increase space for higher level, purposeful thinking is a super power. We also learned that training the brain to be positive and optimistic can promote longevity, success, and happiness. That's powerful. Reframing negativity into positivity and optimism to increase success, happiness, and wellness is another super brain power. We learned in the power of the mind chapter that under certain circumstances belief (either in one's own thoughts or even the placebo effect) can cure the body. That's another great super power. Finding happiness through gratitude and positive feedback is an important super power. Invoking the power of the mind to overcome hardships and to stop harming ourselves is also a super power.

There are several other super powers yet to be covered. This section starts with a chapter on goal setting because it sums up a tangible direction for going forward with all we have learned thus far. Goal setting, planning for the future, and the execution of a plan is a tremendous super power that we have been gifted; these coincide with the development of our frontal lobes.

Beyond these simple day-to-day, super brain powers for good living, there are even greater super brain powers, those available by practicing mindfulness mediation and yoga and tapping into intuition. This is most interesting to me because we need really well-functioning brains to move us out of the devastation we have created on the environment, to dismantle unconscious consumerism, to end the suffering associated with mental illness, to cure diseases, to rise above our animalistic nature, to find peace rather than war, and to adhere to the highest of standards of human and animal rights and social justice. These are big world problems caused mostly by human brains in the first place. My hope is that invoking some super brain powers will help solve some of these problems.

Chapter 8

SMARTER Goals

Sometime after I had finished my PhD and was starting my own business, I was talking with my dad on the phone. I was going on about my long-term plans and all the things I was envisioning and planning for the growth of my business. After listening politely he asked, "Well, do you have it all written down?" I was annoyed at his cutting down my grandiosity, so my cocky response was, "I don't need to write this stuff down! It's in my head and it's very clear up there." His reply was simple: "If you don't write down your goals, you won't achieve them." I, who now own my own business, had earned a PhD, and had won many sports medals, began to rave about all the things I had already achieved without having written any of it down. Clearly, I did not believe that I had to write down goals in order to achieve success. Then my dad knocked me down off my high horse with a wise tongue-in-cheek question: "So, what do you think you'd have achieved if you *had* written down your goals?"

I was stunned and speechless — which my dad enjoyed as he asserted his parenting advice. Although I had great visions and appreciated goal setting, it never occurred to me that I might still be selling myself short. I couldn't help but wonder what else I *could* achieve if I actually wrote down my goals and kept a clear focus on them. The fact that I didn't know made me value the idea of goal setting more than I ever had in the past.

Most of the thinking around goal setting suggests that we need to write down our goals in order to achieve them. At the very least, we need to have them properly articulated in order for us to appropriately move toward them. Even the *law of attraction* would suggest that we clearly express what we desire in order to attract what we desire to us.

The purpose of setting goals is to encourage success, and it is effective. Goal setting is about identifying, exactly and specifically, what outcome is sought, and identifying the means by which the outcome will be achieved. Sounds simple, yet many of us don't take the time to figure out our goals or how to get there. As a result, many of us don't achieve what we hope to simply because we don't have a clear vision of what we are striving for. However, when we do take the time to properly set goals, the exercise can be quite successful. We can achieve our goals efficiently, often more quickly than had we let them float around in our minds haphazardly.

Goal Setting

Keeping Our Goals in Mind and Brain

The areas of the brain that participate in goal-oriented behaviour include the *anterior cingulate* and the *medial prefrontal cortex*. The anterior cingulate is an area of the cortex that holds the awareness of a goal in mind, so to speak; whereas, the medial prefrontal cortex is involved in self-monitoring and keeps track of how we are doing in reference to that goal. For example, if our goal is to earn a gold medal, the anterior cingulate keeps our eye on the prize, and the medial prefrontal cortex keeps track of whether we are on track to achieve it, kind of like a coach keeping us accountable. The involvement of these two areas means that goals involve more than just the desire for a particular outcome. Goals also involve check-ins with how we are doing in pursuit of them. The brain is keeping track of that goal and also how well we are doing in achieving it.

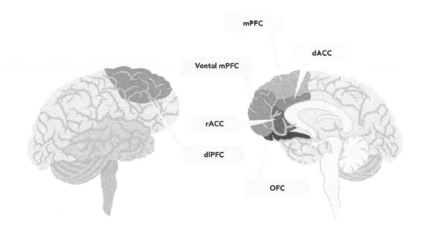

Figure 8. An outside and inside view of the brain's right hemisphere showing 6 cortical regions: dorsal prefrontal cortex (dlPFC), medial (m) PFC, Ventral mPFC, rostral (r) and dorsal (d) anterior cingulate cortex (ACC), and Orbitofrontal Cortex (OFC).

The fact that we have an internal awareness and accountability coach is not too surprising, actually. We need some kind of internal tracker that keeps us focused on our goals. The frontal lobes (where the medial prefrontal cortex is found) are well equipped for that type of high level, executive functioning. My dad had a good point in our conversation about goals. I could *say* that I have a goal, but unless I am actively keeping track of the goal with my medial prefrontal cortex, I am less likely to achieve it. My brain needs to know what goals to keep in mind, making use of my anterior cingulate. My mind and brain see the goals, track them, and direct me to them. Writing them down is one way of clarifying goals for my mind and brain, and there are many smartphone apps available to track goals and progress that, in essence, support the work of the anterior cingulate and the medial prefrontal cortex. With so many distractions in our days, it would seem that the anterior cingulate and medial prefrontal cortex have their work cut out for them. Fortunately, there are many tools we can use to stay aware of and in pursuit of our goals: vision boards, affirmations, calendars, to-do lists, life coaches, friends, and goal-setting charts!

SMART Goals

Keeping our goals in mind and brain can be supported by very clearly defining SMART goals. SMART stands for Specific, Measurable, Accountability and Action, Realistic, and Timed. Consider these two common goals and whether they are SMART: 1) making more money, and 2) losing weight. In fact, neither of those goals is very SMART. What does "more" refer to? How much weight do you want to lose? These goals are neither specific nor measurable. Instead, we might say we want to make $10,000 more or we want to lose 10 pounds. Each of these statements provides a more specific goal.

How will you know if you achieved your goal? This question provokes thinking about whether the goal is measurable. "More" is hard to measure, but measuring weight in pounds, body mass index, waist circumference, or even time spent exercising fulfills the SMART criteria of measurability.

Where's the accountability? Who else knows about our goals? How can we ensure we are somehow accountable? Somehow, we need to have mechanisms in place to ensure we are accountable, whether we put alerts in our calendars to self-check at some future date, have a friend work toward the same goal, or hire a life coach to check in regularly.

We also need to ensure we have an action associated with our goal. A goal without an action might be strong in our mind, but it doesn't allow the goal an opportunity to materialize. If I claim I want to make more money, say that extra $10,000 a year, but change nothing in my lifestyle, then how do I expect to achieve my goal? If instead I look for a part-time job, request a salary-negotiation meeting with my boss, start a side business, or even reduce my spending, these are all actions that could lead to the end goal of having more money. Simply wanting to make more money is not actionable on its own.

Is this goal realistic? Likely, a deeper conversation about successes and failures in the past would be spurred as we discuss whether this goal is not only realistic but realistic for you. Are things that need to be in place before you can make this goal realistic? For example, do you need to gather performance evidence or market comparisons you could bring to the salary-negotiation meeting with your boss? Or perhaps you need a gym membership or a buddy system that will allow you to exercise more and lose that weight. Ask yourself what is needed to be successful.

Being realistic is important, but I like to consider it in relative terms. For people trying to overcome failures of the past, small, easily manageable goals help build confidence and a sense of accomplishment. For others who achieve a lot of success in life, large, optimistic goals might be highly motivating and rewarding. Large goals might even be necessary to prompt progress. It really depends on what is realistic for each person. It's sometimes difficult to assess what's realistic without first attempting to succeed and failing. Setting goals is trial and error for many of us.

By when do you plan to accomplish this goal? Of course, the answer depends on the goal and the person, but nailing down some kind of timeline that feels realistic would help us measure and know when a goal is being achieved. If my goal is to become a millionaire but I have no timestamp on it, then it will be more difficult for me to create an actionable plan to achieve it and won't allow me to easily measure my progress. Timelines are important in that regard.

SMARTER Goals

When I set goals, I don't stop with SMART goals. I strive for SMARTER goals. SMARTER goals are adaptable and involve an Evaluation of change and frequent Reflection. Setting and achieving goals is a process in and of itself, one that is fluid and dynamic and requires reflection as time goes on. What might seem like a reasonable goal today may not be in the future. Blindly holding on to our goals is not ideal. Committing ourselves to our goals, however, is not the same as attaching ourselves to them. Commitment is a motivator. We can still be open to change while being committed. Seeking our goals does not mean that we live only for our future. Our journey toward a goal can and should be a wonderful process irrespective of whether we actually achieve the goal. A great goal is not one that ends with the achievement, but rather a great goal is one that prompts learning and progress and satisfaction along the way. In many ways, the goal might not even matter. When we think of goals in this way, they become SMARTER.

One reason to remain fluid with goals is that, at any given time, something could come into our lives that prevents us from achieving our goals or actually prevents us from needing or wanting that goal any more. For example, if your goal is to achieve freedom at 55, retire early, and enjoy the rest of your life in a beach house with your partner, and if suddenly one of you gets a life-threatening illness related to job stress, you might no longer wish to work that hard for a future that may not come. We all benefit from evaluating and reflecting on our goals and adjusting accordingly. In that case, being attached to a goal or blindly following a goal that no longer serves you could cause you to miss out on what you actually want — a happy life with your partner, not an early retirement!

Being fixated on a goal that no longer makes sense or serves us will steer us in the wrong direction, toward the extrinsic element of the goal — the "I should do this" idea or outcome — instead of remembering the essence or meaning of the goal and enjoying the journey. Therefore, SMARTER goals have all the same goal-setting principles as described above except they include safeguards against this blind spot. If my goal is to be healthy and I decide I want to lose weight, I might do more exercise and weight training. If, as a result, I begin to feel good and healthy, look more toned, and have greater strength, speed, and endurance, then perhaps losing weight is not as important anymore — or may actually be counterproductive as I build more muscle mass and, consequently, gain muscle weight. So we must evaluate and reflect on whether our goals still make sense as originally conceived, and we should repeat this part of the process as often as necessary.

Reflecting frequently, we should be asking ourselves "Why?" Why am I writing this book? Why am I trying out for this team? Why am I building a company? Why do I want to start a family or get married or buy a big house? If we fail to ask and answer these questions, we might lose our purpose and forget about the journey. In fact, that happened to me shortly after finishing my doctorate degree. After spending many years doing graduate work geared toward being a scientist, I went the expected route of doing a post-doctoral fellowship, a necessary step on the way to achieving a tenure-track faculty position at a university. Although there were many elements of my post-doctoral training that I loved, I gradually became quite unhappy in my research lab. I didn't recognize this at first and spent longer than I probably needed to in misery because I failed to sit down and really ask myself why I was pursuing the goal that had motivated me throughout my graduate work. It wasn't until another job opportunity came to me that I began to reflect. In considering this other option, I realized that I could do all the things I really enjoyed about science (intellectual stimulation, learning, creating knowledge, teaching, mentoring, hypothesizing, theorizing, designing studies, and analyzing data) by following a different path. Remembering my "why" through reflection allowed me to make a SMARTER career shift.

SMARTER goals allow us to keep SMART goals in mind, invoke a process of evaluation and reflection, and help us avoid the pitfalls of attachment to our goals and enjoy the process, while still promoting a significant inner and outer sense of accomplishment.

Chapter Assignment

Setting SMART goals is a matter of how we word and frame our goals. Earlier we learned that the way we spoke to ourselves had powerful effects. The same thing happens when we articulate our SMART goals.

Step 1: Start with the big picture.

Ask yourself the following: What is my ultimate vision? What are my goals, dreams, and aspirations? What accomplishments do I hope to achieve? Write these out as a goal.

Step 2: Check to see if your goal is SMARTER.

Specific (Are you specific enough with your goal?):

Measurable (How will you know if you have achieved your goal?):

Accountable (To whom will you be accountable? Yourself? Someone else? A friend? A partner? A colleague?):

Realistic (Can you achieve your goal? If not, can you alter it to be more realistic?):

Timeline (When will you complete your goal?):

Evaluate (At what point will you evaluate this goal to determine whether you are on track to achieve it and whether it still makes sense?):

Reflect (Why are you striving for this goal? Why is it important for you?):

Rewrite your goal if you need to:

Step 3: Break it down.

To really drive this home — and if you seriously want to achieve your goal — you need to start soon and consider how it will really play out. One question to ask is this: What can you do today or tomorrow in pursuit of this goal? Then go forward in time outlining exactly each day, week, month, and/or year as you move toward your goal. Alternatively, you can start at the end and work backwards. I find the latter strategy particularly good for people who are big-picture oriented and possibly a bit grandiose, and those who need a reality check to ensure they can accomplish what they intend to. When you work backwards, it becomes apparent what is and is not realistic. Start by picking the timeline (ten years, five years, or one year from now) and writing out the goal to be achieved by that point. Then ask yourself if this is a SMART goal.

Chapter 9

Mindfulness Meditation and Yoga

Long before I did my first meditation course, I had been fascinated by super powers, always wanting to experience them. My first course taught me, among many things, just how grand our super powers are. I have heard that long-term meditators can do amazing things. In the story of Dipa Ma, author Amy Schmidt claims Ma walks through walls. My friend Anne, who knows Deepak Chopra, also claims that Chopra actually levitates. There are many accounts of super powers arising as a result of serious meditation practice. The idea of super powers is alluring to me even though I recognize that to strive for them goes against the practice of meditation. Nonetheless, as a by-product, they remain interesting and, in my experience, even small super powers arise as a result of heightened attention and awareness.

Mindfulness is the simple act of paying attention to the present moment, and that's what I practised for 10 days straight back in 2007 during my first silent meditation retreat. It was a *Vipassana* course, as taught by S. N. Goenka. *Vipassana* is a Pali (Middle Indo-Aryan language) word that means "to see things are they really are." *Vipassana* is a technique that Goenka claims was passed down directly from the Buddha. Whether this is true or not is kind of like having faith about the Bible being a true account — it may or may not be, and it's for each person to decide. For Goenka's courses (which are the same at hundreds of retreat centres all over the world), students arrive on day 0, leave on day 11, and for the 10 days in between they exist in almost complete silence (see schedule provided to get a sense of what it is like!). Students themselves are not allowed to communicate at all. This includes not talking, writing, reading, making eye contact, and communicating nonverbally in any form. The only speaking is by Goenka himself, during the evening *dharma* (path of enlightenment) talks. These talks are shown on videos recorded in the 80s and are played in each and every course, each and every time you go, all around the world. The schedule is the same every day, in every course, and at every location.

My first *Vipassana* course was intense, and it was one of the most amazing experiences of my life. For the 10 days, my job was simply to observe things as they were in the present moment, without judgment. Without judgment means seeing everything as neither good nor bad, right nor wrong, pain nor pleasure. Everything just is what it is and the goal was to observe the pure sensations and let go of all of the psychological icing we typically place upon experiences, which often manifest as aversion, grasping, or desire. These seem like simple enough instructions, but it takes practice. This is why meditation is often considered a practice while also considered a state of awareness or being.

Vipassana was the most exciting study that I had ever participated in. *I* was the object of my own observation in a rigorous and systematic study. Every day was the same, yet I was changing moment by moment. It was refreshing because after so many years in science, trying to remove the observer bias, it was actually the bias I was studying, first-hand. And who better to study observer bias than the observer herself? As a neuroscientist and psychologist, I gained a whole new depth of understanding about the mind and brain in general and about *my* mind and brain in general. It was neuropsychoidiology in the purest sense.

Indeed, the meditation retreat was one of the hardest things I had ever experienced. But when I returned to the real world, I realized that something very deep had changed inside of me and in my brain. I was calm. So calm! Calmer than I have ever been in my life. Only once before had I experienced something that even approached that level of calmness when I was on antianxiety medication. But after the retreat, as a result of meditation, I was not only calm but also clear-minded. Everything just seemed to make sense. Some semblance of enlightenment, which I had experience in discrete moments in the past, felt more prevalent and more tangible than ever before. I was focused. I was still. I was more comfortable in my skin than ever, symbolically but also technically: the itch of my eczema no longer bothered me, and I felt compassionate and loving toward my body — a body that I had, in the past, hated. I learned to observe sensations as they rose in my mind and body, but not to react to them. I learned to watch these sensations in my mind and body come and go. I learned to notice everything float by like clouds, as my yoga teachers had suggested in the past, but that I had never understood before. My first retreat truly was transformative, mind-altering. It changed the way I saw the world thereafter.

I returned again in 2012. Although both courses were amazing, as was the practice in between, they were completely different experiences. In the first one I experienced my own paranoia, anxiety, and fear of loneliness. During the first course, I experienced my own insanity and fear of insanity. I felt brilliant at times and was devastated that I could not collect my brilliance because I was not permitted to write. I learned to let go as a result. At other times I felt like an idiot, stupider than anyone I had encountered before. In those moments, I was shattered. I fluctuated a lot between both extremes and eventually observed that fluctuation with awe. I fluctuated from high to low, good to bad, sad to elated, and everywhere in between and eventually learned to observe from the middle. The basic principle was that nothing in my life was either of these extremes. I had to try to remove the psychological icing. It was hell, and every day I thought about giving up and leaving. But I didn't. I stayed till the bitter end, and it was amazing too.

Mindfulness, a Practice of Awareness

Mindfulness is the act of paying attention to what is happening now, without imposing judgment on the experience, judgments like bad or good, right or wrong. Mindfulness is pure awareness. Mindfulness is a basic human quality, a way of learning to pay wise attention to whatever is actually happening in our lives rather than focusing on what has happened in the past or what might happen in the future. This allows us to better sense our inward and outward experiences. Mindfulness is about the here and now, not about the then and before. Mindfulness is a practice of cultivating this sense of awareness. Sitting meditation can be mindful, but so can making food, driving, or anything where we are truly in the moment. That concept — being in the moment — has become cliché, so to really appreciate it we do have to practise it. To say it is one thing. To feel it in one's body is a matter of serious practice. Therefore, mindfulness is better understood as an experience.

Raisin-Eating Exercise

Get a big, juicy Thompson raisin. Then spend five to seven minutes eating it. Yes, that long! Look at it. Smell it. Lick it. Place it in your mouth. Move it around your mouth. Let it hit all the parts of your tongue and move it around the inside of your cheek. Let it begin to dissolve and avoid rushing to bite into it. Eat the raisin as if it's the only thing you have to do today. Take more time eating this raisin than you have ever before. Focusing attention in this manner is mindfulness. It's a practice of awareness. In this case, it's awareness of eating a raisin.

Notes:

Mindfulness has been defined by many people, including Dr. Jon Kabat-Zinn, the founder and scientist behind the secular health-related program called Mindfulness-Based Stress Reduction (or MBSR). He defines mindfulness as the act of paying attention to the present moment without judgment.[54] Mindfulness, as is done in MBSR courses, is based on the *Vipassana* technique of meditation, where students learn the practice of awareness of body sensation.

Mindfulness, according to neurologist Dr. Daniel Siegel, offers a way of being aware that can serve as a gateway toward a more vital mode of being in the world. In other words, Siegel suggests that we become attuned to ourselves. He describes mindfulness as a form of mindful awareness and as a form of self-relationship, an internal form of attunement that creates emotional longevity, helping us achieve states of well-being and health. Ultimately the practices that develop mindful ways of being enable the individual to perceive the deeper nature of how the mind functions. There are many ways to cultivate mindful awareness. Mindfulness meditation is thought to be especially important for training attention and letting go of a strict identification with the activities of the mind as being the full identity of the individual.

According to monk Thich Nhat Hanh,[55] mindfulness is the energy of being aware and awake to the present moment. It is the continuous practice of touching life deeply in every moment of daily life. To be mindful is to be truly alive, present, and at one with those around you and with what you are doing. We bring our bodies and

---

54. Jon Kabat-Zinn, *Full Catastrophe Living: Using the Wisdom of Your Body and Mind to Face Stress, Pain, and Illness* (New York: Bantam Dell, 1990).

55. Monk Thich Nhat Hanh founded Plum Village (see plumvillage.org), a Buddhist monastery that hosts meditation retreats.

minds into harmony while we wash the dishes, drive the car, or take a morning shower.

Another way of considering mindfulness is as a deconditioning. We enter this life reliant on our caregivers and our instincts, and then over time we begin to develop a sense of self and a sense of agency. From a Western psychology perspective, we spend our lives trying to achieve healthy self-esteem and, possibly, self-actualization. But from an Eastern psychology perspective, for example, yoga or Buddhism, we spend our lives as a practice trying to detach from all that makes up our identities and from the contents of our minds. This is the dissolving of the self as described in the chapter on self.

Kabat-Zinn refers to seven pillars of mindfulness, which I think are helpful anchors when we consider applying mindfulness to our daily lives: 1) non-judgment, 2) patience, 3) a beginner's mind, 4) trust, 5) non-striving, 6) acceptance, and 7) letting go. In studying these concepts, and mindfulness in general, we should be wary of not just coming to know them from an intellectual perspective, but learning these from an experiential perspective, from an embodied perspective. In fact, in all of the teachings, practice and study are both important elements in the eventual reduction of suffering or enlightenment.

Mindful Teeth-Brushing Exercise

Next time you brush your teeth in the morning, do it mindfully. Start with actually staying in the bathroom, which is challenging enough for some of us who wander around trying to get other things ready in the morning. Stay put and focus all of your attention on brushing your teeth. Notice how the bristles move across your teeth. Feel the foam from the toothpaste growing in your mouth. And every time you notice your mind wandering, invite your attention back to the object of your focus: brushing your teeth. Do this repeatedly for two weeks and notice how you feel and how the quality of your awareness and attention changes over the two weeks.

Notes:

Most people report that they develop a sense of calmness as they brush their teeth, and that they can keep their attention focused more by the end of the two weeks. This exercise can be repeated with any number of activities, such as showering, making dinner, or driving.

There are many ways we can practise mindfulness because, of course, it really is just paying attention to the present moment without judgment. Any and every moment is an opportunity to pay attention, and therefore an opportunity to practise mindfulness.

Mindfulness Meditation

Many of us think of meditation as sitting cross-legged with eyes closed and blissing out. In fact, this is not what meditation is. Blissful experiences are by-products of a practice. Meditation is a state of mind and a process. To fully understand meditation, one must practise. With continual practice we come to more deeply understand mediation from an internal perspective. Meditation is *not* about having a still mind per se, it's about *developing* a still mind or, rather, observing the mind as still. A common experience among new students to meditation (or yoga for that matter) is a feeling of "I can't do this." "I can't do this because my mind is too scattered," says the new meditation student. "I can't do this because I'm not flexible enough," says the new yoga student. "That's why you practise," responds the teacher. Meditation (or yoga) is a journey and the goal.

What does a "still mind" mean? Perhaps it's easier to recognize what a non-still mind is like. Grab an alarm, set it for five minutes, close your eyes, and then try for to focus on just your breath without losing your focus. Just focus on your breath and nothing else. If any thoughts arise, let them go. Start now.

Sitting Body Scan Meditation

Find a comfortable sitting position. It doesn't matter how you sit; just try to be comfortable and sit with a straight spine. Begin by focusing on your breath. Pay attention to how your breath moves in and out of your body, over and over again. Notice the sensations around your nose, if you are breathing through your nose, or around your mouth, if you are breathing through your mouth. After a few minutes of that, allow your attention to move to the top of your head. Notice if there are any sensations there. Try not to move your attention away even if you feel nothing at first. Keep focused for a bit, then gently move your attention down your head, and scan all around your head, spending enough time to detect any sensations that might be present. Keep moving your attention systematically downward to your face, your neck, your shoulders, your chest, your back, and so on. Move in a pattern down toward to your feet, then come back to the top of your head, and repeat. Continue this for however long you want. Try starting with 10 minutes. Notice how you feel afterwards. Notice how you feel after several sessions.

Notes:

At first most people find meditation challenging. Keep moving your attention systematically. Eventually, the more you practise, the easier you will find it. Most people find it calming and relaxing, and they begin to notice sensations they never had before. This is a great way of developing mind-body awareness and a mind-body connection. Often we forget to pay attention to these experiences, and we can go days or even months without feeling anything in our bodies. That might explain why we slump over at our desks or computers and end up with body aches and pains.

Such a brain-training practice offers many health benefits, particularly those associated with high stress. Spiritually though, mindfulness also offers a path to pure awareness, or nirvana, or enlightenment, or seeing things as they truly are. This takes the mind to a whole other level, and in my small experience with mindfulness practice, it totally changed how I saw the world and myself in it.

Meditation is practising being in the present moment. It is about focusing our attention on the breath, quietness, taming the scattered mind, cultivating open awareness not based on thinking, and ultimately finding calmness and being peaceful. It's also the practice of all of that. It's a journey and a goal. That's hard for our Western minds to wrap around. Indeed, meditation is about getting out of our minds and not being what our mind produces (thoughts, reactions, labels, anxieties, and so on), but just being. In many ways meditation is about being (with) our nervous systems and sensing what is in this life, from physical sensations on our skin to observing the thoughts that go through our minds, but without *becoming* our thoughts. I know it seems that I am making this complicated. It is complicated if we try to understand it with our minds. But if we just experience meditation, it is quite simple. It is pure awareness. Anyone can do it.

Yoga, a Mindfulness Meditation Practice

My first experience with yoga was a workout video I did in 1998 with my roommate. She and I were competitive athletes, not just with each other but generally. The first time we did the video, I was sore for a week because we each tried to hold our downward-facing dog longer than the other. We, like many people, believed yoga was about endurance and muscling through, contorting as much as we could. It's not. That aspect of yoga is indulging the ego rather than attempting to transcend it. But my roommate and I didn't know that at the time.

Learning that yoga is not about sweating, exercise, endurance, or competitiveness and learning to let go of the ego can take years, if not a lifetime, to grasp. My ego still joins me on my yoga mat often enough, like when I judge whether I am working hard enough in my own practice or if my body looks the way I want it to, or if I'm in a class and I judge the teacher's choice of words. Sometimes my ego wants to "win" at yoga, but at other times my ego lets go and I fall into a space where I observe my mind's reactions to the postures and my relationship with hardship. I have to ask, is this too much or too little? What is the nature and experience of my body in this moment?

In *Kripalu* yoga (the tradition in which I am trained) we call the space between too much and too little the "edge."

If you want to do a simple exercise to experience this edge, try pulling your little finger backwards. Try to observe the moment when you just begin to feel the stretch. That's the beginning of your edge. Then, with awareness, keep pulling your finger back to a point where it feels really uncomfortable. That next point is an upper version of your edge, but not the end. You would be surprised how far you could go without damaging yourself. You can also try this kneeling down with your hips/bum resting on your ankles. Let your toes curl under to prop you up a bit. It will most likely feel tight, and eventually it may even feel excruciating. But you can do it if you pay deep attention to those sensations. Of course, there is an upper edge there, but most of us don't know where that is. Within the exploration lies a super brain power, hidden to most of us because we have not had to use it. Well-trained athletes, people who can walk on fire, people who can walk through doors, and very well-trained meditators and yogis know versions of it.

Yoga is about awareness, plain and simple, just like mindfulness and *Vipassana* meditation. As we observe our bodies moving through space in stillness, there are many things to take notice of. Intensity? What is the nature of this intensity, and how am I reacting to it? Boredom? What is the nature of this boredom in a posture, and how am I reacting to it? Fear? What is the nature of fear? Identity? Do I feel good about what I can accomplish? These are all wonderful ways yoga becomes a study of the mind-body relationship, or becomes a meditative or mindfulness practice itself. "I don't like this." "I do like this." "I am bored." "I am strong." All of these are judgments beyond the moment of pure sensation and awareness, just some of the judgments that take away from the moment in its pure form. They are part of what we learn to let go of as we continue to practise meditation, mindfulness, and yoga.

Benefits of Mindfulness

Our brains are responding to the experiences around us when we are pushed to multitask or switch focus. We all have shortened attention spans. We spend hours a day flipping from topic to topic, task to task, conversation to conversation, kid to kid, job to job. We are essentially training our brains to be scattered. Flipping from thought to thought becomes the norm, and our brains become wired to support that, whether we like it or not. The result is that staying on task becomes more difficult. Our brains are not primed to stay on task. They flip more efficiently. In fact, the average attention span seems to have decreased, and some claim that change is reflected in films being produced. For example, the average shot length of an English-language film has declined from about 12 seconds in 1930 to about 2.5 seconds today, according to researcher James Cutting at Cornell University.[56] We are inadvertently training our brains to be scattered because scattered focus is what we do a lot of. The brain rewires itself to this new baseline state.

One of the reasons that meditation is so beneficial is because it resets our brains to a new baseline of attention, focus, and awareness. We are retraining the

---

56. James Cutting and Catalina Iricinschi, "Re-Presentations of Space in Hollywood Movies," *Cognitive Science* 39 (2015): 434–456.

brain to be single-focused, attentive, and aware of inner and outer experiences. These experiences are often ignored when we inadvertently fail to pay enough attention to them, when instead we are at the whim of a mind tethered to very little, haphazardly roaming from task to task. But with training (or practice) on being focused, we can learn to control the mind and keep it focused on what we intend to focus on. That is a brain super power in today's world, no doubt!

Being able to focus on one thing for a longer period of time inevitably brings about a sense of calmness and reduces stress because we are less susceptible to the penetrating thoughts of the future or the past. Worries over what will be or regrets of the past become much less, if at all, intrusive. By focusing on what is happening now, we are not threatened by what could potentially happen in the future or what already happened in the past. We are "in the now," as it is often referred to. The "now" allows the brain not to be stuck in worry mode, in reference to either the past (like "why did I do something so stupid" or "I wish I had followed that advice") or the future ("when am I going to get groceries" or "how will I ever meet that deadline"). Thinking about the past and the future represents an entirely different brain activation than paying attention to what is going on now, which is largely sensation-based. I am breathing. I feel warm. I feel tingly. I feel a lump in my throat. Being worried, scatter-brained, or pessimistic are just some of the ways we reinforce our brain to think a certain way, and these thought patterns tend to have negative consequences on our physical and mental health. The good news is that if we changed the brain to think in one way, we can probably change it to think another way. And that's what meditation teaches us.

There is a great deal of research on the health benefits of mindfulness right now, with many age-old claims being substantiated in the scientific world. Mindfulness has been shown to be beneficial in stress, addictions, generalized anxiety, depression, chronic pain, cancer, cardiac disease, burnout, multiple sclerosis, and psoriasis, and the list is growing.

Research specifically on meditators (both short-term and long-term) has also generated scientific evidence showing a greater sense of well-being. For example, long-term meditators have more feelings of happiness than short-term meditators.[57] This greater happiness shows up with greater neural activity in the left frontal part of the brain, an area known to be related to positive emotion. In fact, the frontal region of the brain is well documented in science to be involved in meditative states, which likely has to do with that region's role in cognitive functions like attention, inhibition of socially unacceptable behaviours, and retention of something in memory (or mind).

The benefits of yoga are also vast. For example, yoga has been shown to help with burnout, self care, cardiovascular issues in seniors, headaches, multiple sclerosis, depressive symptoms, and several improvements for individuals with PTSD, such as sleep, positive emotion, perceived stress, anxiety, and resilience. Yoga has also been shown to reduce oxidative stress and improve antioxidant levels while also reducing stress

---

57. Matthieu Ricard, Antoine Lutz, and Richard R. Davidson, "Mind of the Meditator: Contemplative Practices that Extend Back Thousands of Years How a Multitude of Benefits for Both Body and Mind," *Scientific American* 311, no.5 (2014): 38–45.

hormones and improving immune function. This list is by no means exhaustive.[58]

Earlier studies reported that those who follow a yoga program are at a decreased risk of being diagnosed with an eating disorder[59] and that a yoga-at-work program resulted in individuals feeling more clear-minded, composed, elated, and energetic, with a greater sense of life purpose and satisfaction, and greater self-confidence in stressful situations.[60] Another 12-week yoga program found women to have increased sexual desire, arousal, lubrication, orgasm, satisfaction, and reduced pain, and showed men to have increased desire, satisfaction, performance, confidence, partner synchronization, erection, ejaculatory control, and orgasm.[61,62]

Why do all of these benefits arise? Some aspects of yoga and meditation overlap, particularly the stress-reduction aspect. It is simply not healthy for our brains and minds to be anywhere but in the present moment for extended periods of time, and both meditation and yoga yoke our minds with our bodies. When we are not "here," we fail to pay attention to important health signals as they arise, or we directly cause havoc in our bodies. Whereas,

---

58. For a review of the benefits of yoga, type "yoga and benefits" into the U.S. National Library of Medicine's PubMed (http://www.ncbi.nlm.nih.gov/pubmed/?term=yoga+benefits)

59. Jessalyn Klein and Catherine Cook-Cottone, "The Effects of Yoga on Eating Disorder Symptoms and Correlates: A Review," *International Journal of Yoga Therapy* 23, no. 2 (2013): 41–50.

60. Ned Hartfiel et al., "The Effectiveness of Yoga for the Improvement of Well-Being and Resilience to Stress in the Workplace," *Scandinavian Journal of Work, Environment & Health* 37, no. 1 (2011): 70–6.

61. Vikas Dhikav et al., "Yoga in Male Sexual Functioning: A Noncomparative Pilot Study," *The Journal of Sexual Medicine* 7, no. 10 (2010): 3460–6.

62. Vikas Dhikav et al., "Yoga in Female Sexual Functions," *The Journal of Sexual Medicine* 7, no. 2pt2, (2009): 964–70.

paying attention to the present moment seems to be of great benefit to both physical and mental health. The physical element of yoga has an added value of allowing the body to refrain from stagnation, to stretch, and to literally move in ways that promote health.

The Neuroscience of Meditation and Yoga

In 2005 I was fortunate to be in the seventh row of a packed auditorium at the Society for Neuroscience annual meeting. Each year the meeting draws over 30,000 neuroscientists, and that year the Dalai Lama spoke during one of the keynote lectures. I was so excited to hear this momentous lecture. I don't remember all that he said, but one thing stood out quite clearly: he urged the neuroscientists to study meditation as much as possible because if it helped bring meditation to the world and helped ease suffering, then it was a good thing. He condoned the scientific study of meditation, very optimistically, claiming that he would revise this thinking if science provided new insights. From what I have read, he is scientifically literate enough to understand how difficult it would be to fully capture the essence of meditation in a laboratory setting, but he nonetheless valued scientific contributions.

Since that address there has, indeed, been a surge of research on the neuroscience of mediation, much more than on the neuroscience of yoga. Even before the Dalai Lama's address, in 2004 one of the neuroscience-meditation research gurus, Richard Davidson, published a seminal paper where he noted that a rare form of high-frequency brain waves (gamma waves) were much more prominent in long-term meditators compared to people new to meditation.[63] Gamma waves are of interest because they seem to represent several aspects of cognition, including working memory, selective attention, and conscious perception of the external world. We have continued to see these gamma waves emerge as a reliable feature of meditation. In addition, other brainwave changes have also been noticed, for example, with theta (characteristic of learning) and alpha (characteristic of relaxed wakefulness with eyes closed) waves. These brainwaves also seem to relate to the brain's ability to focus and to reduce distracting stimuli in our environment. Given that these are the same experiences noticed by meditators, it is not surprising that the brain shows corresponding brainwaves.

In a *Scientific American* article[64] published in November 2014, neuroscience-meditation gurus Matthieu Ricard, Antoine Lutz, and Richard Davidson provide an excellent overview of how different types of meditative practices show up in the brain, including mindfulness meditation. For example, in mindfulness meditation, expert meditators have decreased activity in the amygdala and insular cortex, two areas involved in emotional behaviours like anxiety. In people who are practiced at focusing their attention (like monks), the dorsolateral prefrontal cortex stays active, reflecting the height of mindfulness and sustained attention. However, even in those experienced meditators, attention can be lost, and at that time the brain system referred to as the "default-

63. Antoine Lutz et al., "Long-Term Meditators Self-Induce High-Amplitude Gamma Synchrony During Mental Practice," *PNAS* 101, no. 46 (2004): 16369–16373.

64. Ricard, Lutz, and Davidson, "Mind of the Meditator," 38–45.

mode network," which includes the precuneus and the posterior cingulate cortex, is activated. Once aware that the mind has wandered, the "salience network" kicks in, involving the anterior insula and anterior cingulate cortex. Upon awareness of a distraction, a reorientation of focus can happen, and that involves activation of the dorsolateral prefrontal cortex again along with the inferior parietal cortex. These areas seem to help disengage our minds from the distraction and then allow the mind to refocus, at which point only the dorsolateral cortex remains active.

Meditation also changes our neurochemistry and neuroanatomy. For example, meditation is associated with changes in dopamine[65,66] (our reward chemical), serotonin[67] (the "happiness and wellness" chemical that antidepressant medications aim to increase), and melatonin[68] (our sleep-well, be-well chemical that we often don't get enough of in winter). Meditation has also been related to a variety of anatomical changes in the brain. For example, people tend to experience reduced stress with meditation, and this effect correlates with decreased cells[69]

65. Ulrich Kirk and P. Read Montague, "Mindfulness Meditation Modulates Reward Prediction Errors in a Passive Conditioning Task," *Frontiers in Neuroscience* 6, no. 90 (2015), doi: 10.3389/fpsyg.2015.00090

66. Ye-Ha Jung et al., "The Effects of Mind-Body Training on Stress Reduction, Positive Affect, and Plasma Catecholamines," *Neuroscience Letters* 479, no. 2 (2010): 138–142.

67. Xinjun Yu et al., "Activation of the Anterior Prefrontal Cortex and Serotonergic System is Associated with Improvements in Mood and EEG Changes Induced by Zen Meditation Practice in Novices," *International Journal of Psychophysiology* 80, no. 2 (2011): 103–11.

68. Chien-Hui Liou et al., "Detection of Nighttime Melatonin Level in Chinese Original Quiet Sitting," *Journal of the Formosan Medical Association* 109, no. 10 (2010): 694–701.

69. Specifically, it was a decrease in grey matter. But the brain consists primarily of grey matter and white matter, anatomically speaking. White

in the amygdala,[70] the area of the brain involved in stress, fear, and anxiety. The amygdala is involved in interpreting things around us as dangerous. If there is a decrease in density in that region, the individual may become less responsive to stressors. Other grey matter results were also found within the hippocampus, possibly reflecting a heightened awareness of our surroundings.[71]

In another interesting study, researchers explored the idea that our relationship to the sensations in our body changes during meditation. In this study, researchers were examining the responses of experienced meditators to painful experiences imposed upon them (for example, an arm immersed in an ice bath). They found reduced activity in the insula, cingulate, and amygdala just *before* pain was experienced, but heightened activity *during* pain.[72] The brain activation corresponded with reports of the reduced pain. When asked what the participants were doing in the study during the ice bath immersion, the meditators said they were experiencing the sensations much like they would do during any meditation practice — by paying attention. This idea of paying attention to the

matter is the myelination of the neuron axons, and relates to speed of conduction (that is, how quickly we can send the message from one neuron to the other). Grey matter is composed of cell bodies, the parts of the brain that ultimately add up all of the chemical messages from the neighbouring neurons. The cell body also contains lots of machinery to direct the neuron's activities.

70. The basolaternal amygdala. Britta K. Hölzel et al., "Stress Reduction Correlates with Structural Changes in the Amygdala," *Social Cognition and Affective Neuroscience* 5, no. 1 (2009): 11–17.

71. Specifically, the subiculum. Eileen Luders et al., "Meditation Effects within the Hippocampal Complex Revealed by Voxel-Based Morphometry and Cytoarchitectonic Probabilistic Mapping," *Frontiers in Psychology* 4, no. 398 (2013) doi: 10.3389/fpsyg.2013.00398

72. Antoine Lutz et al., "Altered Anterior Insula Activation during Anticipation and Experience of Painful Stimuli in Expert Meditators," *Neuroimage* 64 (2013): 538–46.

pain is a significant deviation from how we generally think of pain. Normally, we try to avoid it, but mindfulness teaches us to pay attention to it but only to the sensations, without judging those sensations as good or bad. The judgment is thought to contribute to our suffering or, as referred to here, "pain." I can attest to a similar experience. During my first 10-day meditation course, I began with an incredible amount of pain in my hips. But after three days the pain transformed. It was not that I became immune or habituated to the pain or that the pain went away but, rather, my relationship to the sensations changed such that I no longer experienced it as pain.

Since 2005 there has been a huge focus on the neuroscience of meditation. More and more is learned about the effects of meditation each day, largely spurred by the interest of the Dalai Lama and others. This is exciting, indeed, but those of us who have done any intense meditation have already had some incredible experiences, and we already know there are intense things changing in the brain as a result of meditation. Meditation changed my brain by decreasing my sensitivity to pain, itches, and general discomfort in my body. I suspect that I had generally reduced activity in the insula, cingulate, and amygdala, and then when some sensation did arise in my body, I experienced it as it was, which likely caused activity in those areas to increase. The experiences I felt, however, were not of heightened pain or discomfort. Rather, they were experiences of seeing things as they are. This is an important skill as we proceed through life trying to rid ourselves of delusions, trying to know our true selves, and trying to use our brains to better sense and perceive reality. I also experienced an incredible change in my anxiety levels, which I highly suspect corresponded to changes within my serotonergic system and amygdala. One other important thing I came away from my first meditation course with was a reduction in cravings. I felt no cravings for sugar, for food, or for consumerism, likely as a result of changes in dopamine levels.

During my practice, I have felt deep shifts in sensation, perception, attention, awareness, emotions, thinking, and behaviour, all brain-based psychological experiences. This higher level of understanding instilled in me a faith that the practice of meditation was doing something quite powerful to my sense of self, my mind, my body, and my brain even if neuroscientists do not (yet) understand all the pieces. In any case, both science and my own personal experience tells me that something powerful happens in the practice of meditation. And again, it's best for us each to experience it for ourselves. My second 10-day course was vastly different from my first. It felt much less transformational. Yet still, I know deeply that the experience continued to impact my brain.

Yoga, on the other hand, has had much less attention in the neuroscience world although it has been shown to affect many cognitive experiences for elderly individuals such as improving immediate and delayed recall, verbal and visual memory, attention, working memory, verbal fluency, executive function and processing speed.[73] Yoga has also been shown to exert psychological benefits for non-elderly, including improved depressive symptoms,

---

73. Ankur Sachdeva, Kuldip Kumar, and Kuljeet Singh Anand, "Non Pharmacological Cognitive Enhancers – Current Perspective," *Journal of Clinical and Diagnostic Research* 9, no. 7 (2015): VE01–VE06.

stress reduction, emotional regulation, and general improvement in well-being.[74]

A study published in 2014[75] reports some interesting brain changes. The participants in this study were premenopausal women who practiced yoga three times a week for twelve weeks compared to women who had not practiced. Those in the yoga group showed steady-state levels of serotonin compared to the control group, in which a drop-off was observed. Therefore, yoga seems to offer protection against the tendency for our levels to drop. On the other hand, those in the yoga group had increased levels of a brain chemical known as BDNF, or brain-derived neurotrophic factor. BDNF is an interesting factor because in some ways it's the hallmark of neuroplasticity. BNDF is necessary for neurogenesis and for cognition functions like memory. Without it, we don't do so well.

Yoga has also been shown to increase cells in the hippocampus,[76] similar to the effect we see with meditation. In fact, the number of hours of weekly practice correlates with changes in several brain areas including somatosensory cortex, precuneus/posterior cingulate cortex, hippocampus, and primary visual cortex,[77] such that an increase in practice was associated with an increase in brain tissue in those regions. Each of

---

74. Ibid.

75. Moseon Lee, Woongjoon Moon, and Jaehee Kim, "Effect of Yoga on Pain, Brain-Derived Neurotrophic Factor, And Serotonin in Premenopausal Women with Chronic Low Back Pain," *Evidence-Based Complementary and Alternative Medicine*, Article ID 203173 (2014): doi: 10.1155/2014/203173.

76. V. R. Hariprasad et al., "Yoga Increases the Volume of the Hippocampus in Elderly Subjects," *Indian Journal of Psychiatry* 55, Suppl 3 (2013): S394–6.

77. Chantal Villemure et al., "Neuroprotective Effects of Yoga Practice: Age-, Experience-, and Frequency-Dependent Plasticity," *Frontiers in Human Neuroscience* 9, no. 281 (2015): doi: 10.3389/fnhum.2015.00281.

these areas is associated with awareness in some form or another. When I think about my own personal practice, I can attest to having gained a significant amount of awareness of my body, so much so that I believe it has helped me not only take care of my body but also use it more wisely. In 2008 I completed my yoga teacher training and was engaging in many hours of yoga practice a week. Just before that, I had been on the verge of quitting ultimate Frisbee because it was so hard on my body. I was 32 at the time, and I just assumed I was nearing the end of my athletic life. Beginning a deeper practice of yoga completely changed my experience in my own body. I became much more compassionate toward it, cared for it better, and "listened" to it in a way that finally made sense to me. As a result, I continued to play ultimate and went on to spend the years between 33 and 38 feeling like I was in the best shape of my life. Even now, I feel like my body will last much longer if I continue to practise yoga. Cognitively and emotionally, I also feel much more balanced, calm, loving, and generally well when I am actively doing my yoga practice.

Mindfulness Meditation and Yoga Super Brain Powers

A scientific paper published in 2014 in the journal *Frontiers in Aging Neuroscience*[78] showed that yoga and meditation slowed the decline of fluid intelligence as we age. Fluid intelligence involves the ability to detect novel

---

78. Tim Gard et al., "Fluid Intelligence and Brain Functional Organization in Aging Yoga and Meditation Practitioners," *Frontiers in Aging Neuroscience* 6, no. 1 (2014): 76, doi:10.3389/fnagi.2014.00076.

patterns, extrapolate, solve problems, and use something other than knowledge (like wisdom) to think. The influence of yoga and meditation on that human cognitive capacity strikes me as a very clear super brain power.

When we still our minds, limit our low-quality thinking, and start to experience ourselves at a more basic level of pure sensation, we start to tap into experiences we didn't even know were there. Our relationship with our existence changes, not because we endure this existence but because we begin to see ourselves as the pure sensation that makes up our experiences, over and above the mental reactions and conditionings that we inherit unconsciously. Mindfulness meditation and yoga allow us to challenge the notion that everything is exactly how we thought it was. The new ways of seeing, thinking, and behaving become engrained in our new brain architecture and set us on a new path of awareness and discovery, with experiences that were otherwise not available. Until the brain is able to receive, sense, and perceive differently, we are destined to continue on seeing things as they have always been.

During my second meditation course I gained insight into this phenomenon. As I became more highly attuned to my sensations, I also felt attuned to sensations around me, those that extended beyond my body. Things began to change form. The moon, the trees, and the earth all appeared different. I began to wonder if a change in sensation and a change in perception is what allowed Dipa Ma to walk through walls. Maybe she saw openings that other people couldn't see. Maybe she wasn't actually walking through walls, but rather she saw the places where the perceived solid was not solid at all.

Walking through walls is likely unbelievable to most people, myself once included. I am skeptical indeed. But my friend who claims to have witnessed Deepak Chopra levitating is a well-educated doctor of philosophy who teaches at a university. I have to believe she believes she saw this. Is it because he actually levitated? Is it because lifting the body above the earth had taken on a whole new meaning? Or is it because there is something there that we simply don't see with our unaware eyes and mind? I don't know. But I am curious.

In another example, I recall lying in bed at my meditation course after a long day of practice. I felt my body there, still in preparation for going to sleep. Then, all of sudden, I felt an incredible sense of lightness come over me. I felt no different than the air around me. The sensations on my body became an extension of the atmosphere around me. And then something began to lift away. My sense of "me" began to rise, floating up to the ceiling. For that moment I understood, from a very experiential perspective, the true essence of non-attachment as I let my energy drift away from my body. And then I thought about what that might really mean to let go of my body, of my mind, and of myself. My heart began to race. I panicked. My body became real again. Very real! I became afraid of letting go of my body. I was terrified of experiencing the oneness of the universe. With that, I was back. Oneness, non-attachment, letting go... all out the window and there I lay, feeling the rush of myself in my ever-sensing body. In hindsight, I wonder if that is how Dipa Ma can walk through walls.

I can't help but think that this fear of letting go — of what I know of my body, of what I have come to rely on as truths of how we exist, or of the laws that govern our rationality — is present in all of us who fail to experience our super brain powers. We are blind to what's possible because our way of seeing the world is narrowed by previous experience, by the need for security, and by deep fear of the unknown. At the very least, I know those fears are true for me, despite an opposing motivation to seek. This is classic approach-avoidant behaviour manifested in a search for a bigger truth.

In many yoga and other Eastern psychology texts, psychic powers are described. How do I reconcile these as a skeptical neuroscientist who has not experienced these super powers? I don't. In fact, I am reminded of what drew me to psychology in the first place — psychic abilities and other experiences we collect under the term "parapsychology." As a trained scientist, I am very well aware that we do not know the full extent of what the brain is capable of. We do not know all the ways in which the brain communicates and functions. We are still righting scientific fallacies in neuroscience of the past, and we will continue to revamp our theories of today. Neuroscience does not have all the answers. Science does not have all the answers. Any one brain does not have all the answers. But I am curious about how heightened awareness of my own sensations might evolve into a super power, and I'm quite confident that it can, mostly because I have already experienced some evidence in support of that possibility.

Finally, I will end with the super power of happiness. Psychologist Dr. Daniel Gilbert has discovered that a wandering mind is not a happy mind. A few years ago, Gilbert published a book for the general market, called

Stumbling on Happiness,[79] where he describes some of the tricks of the mind and how they can both create and impede happiness. More recently, Gilbert and his student Matthew Killingsworth published a paper in one of the top scientific journals, *Science*,[80] for which they tracked people's happiness at random intervals during the day. People in the study would get random notifications to their smartphones prompting them to immediately stop what they were doing to fill out a survey capturing what they were doing, thinking, and feeling. What Gilbert and Killingsworth first found was that the frequency of mind wandering in people was pretty high: almost half of the samples involved minds that were not focused on what they were supposed to be doing. The wandering-mind phenomenon is nothing new to scientists (or to humans!), but the percentage they found in this study (47 per cent) was much higher than what previous scientists had found in laboratory settings, owing to the new smartphone technology that allowed for the data to be collected in real-life settings. More interesting, however, was that when people's minds were wandering, they were less happy than when their minds were focused, regardless of whether they were involved in enjoyable activities or not. This wandering mind is essentially the opposite of the focused mind that arises with the practice of mindfulness.

---

79. N.B. as far as science is concerned, the science in the book is becoming a bit outdated, but it serves as a great review of what existed before it was published. It is still worthwhile for someone who really cares to learn about this field.

80. Matthew A. Killingsworth and Daniel T. Gilbert, "A Wandering Mind is an Unhappy Mind," *Science* 330, no. 6006 (2010): 932.

Developing Mindfulness Meditation and Yoga Super Powers

I have experienced profound effects with each of these practices, but I hesitate to offer them as a cure-all although I believe them to be. I learned when, several years ago, I encouraged a client to do the same 10-day silent meditation course that I had done. I knew how much I benefited from it and thought he might benefit too, and he seemed interested in it at the time. I'm sure he learned something, but it must have happened within the first two days because after that he split. Literally, he ran away from the centre to the nearest town, jumped on a bus, and then flew home!

Insight meditation teacher Jack Kornfield spoke about the desire to cite meditation as a cure-all when he returned to his clinical psychology practice after years as a monk, a journey he discussed in one of his books, *A Path with Heart.* He quickly learned that some people needed first to be grounded in their own bodies before they could actually attempt to transcend and liberate themselves and their minds. In such cases (for example, in people with schizophrenia), he believed more physical practices like yoga were helpful. That said, I have also known people who reacted poorly to yoga, both when it involved the elements beyond a physical fitness practice and when it was engaged in as a physical practice without the teacher laying a foundation for body awareness. Physical and mental damage can occur.

What I learned from each of these experiences is that mindfulness meditation and yoga are not benign, which is part of what makes them so enriching and beautiful as practices. But under the wrong conditions they can be damaging. I recommend that each of these practices be pursued with the guidance of a well-practiced and knowledgeable teacher. That's not to say that a teacher has all the answers, but the teacher is there to serve as a guide throughout the process. Whether we are beginners or relatively experienced, teachers are important because they have walked the path themselves and can support common encounters along their students' paths.

It's also worth noting that mindfulness meditation and yoga are being taught by many people all over the world, some of whom are more capable of teaching than others. Some are earning certifications and some are not. Unfortunately, I don't think "certification" is the best judge of this sort of thing because certification has no way of accounting for how many lives a person has spent studying these techniques and philosophies before this one, or how invested any one person's heart has been in studying and learning. Some teachings will be good, some will be bad, some might be helpful, and some might even be harmful. My hope is that my offerings here are more helpful than harmful. I do feel it's important to discuss mindfulness meditation and yoga here because these practices are transformative and should not be withheld from people.

For me, yoga has been a saviour. It has taught me a new lifestyle, one that involves self-compassion, self-love, and body awareness. I remember a moment on my yoga mat when I was reaching forward trying to touch my toes, but pain was radiating up the back of my legs and through my hips. I couldn't go very far and I kept thinking, "I can't be a yoga teacher if I can't touch my toes!" I was looking around and seeing my classmates able to touch their toes, and it only made me want to reach farther and push deeper through the pain. I heard a voice inside say, "What's wrong with you? Work harder. Be better, stronger, faster, slimmer..." It was a familiar voice. It was a familiar pain. Then something shifted. I became an observer for a moment and saw what I was doing. I was fortunate enough to wake up in that moment on my yoga mat, and I began to see the terribly abusive relationship I had with myself. It's an abuse many of us know well. We hate, hit, distort, and hide our bodies or parts of them. We yell, scream, and verbally assault ourselves. We speak to and think of ourselves in ways we would never do to another person, yet we condone it toward ourselves. I was fortunate to wake up one moment on my yoga mat to my own self-abuse. It was the moment when I started a new relationship with myself and my body, and then the beauty of my body began to emerge in my eyes, finally. Practical changes occurred too. I learned to listen to my body when it told me it hurt and needed a rest or a break. I began to listen to sensations that had emerged for a long time. This listening allows me to be strong and healthy and wise at my age, more able to run, bike, play ultimate, and do all the physical things that once caused me so much pain. Meditation has also been instrumental in my path of personal development, wellness, enlightenment, and letting go, or however you want to refer to it. I know, without a doubt, that I have been transformed for the better through my practice.

Chapter Assignment: Mindfulness Practice

I do encourage people to practise whatever form of mindfulness meditation and yoga they can in whatever capacity they can. Perhaps it is just a morning start with three deep breaths. Perhaps it is limited to wearing yoga clothing to feel better inside. Perhaps it is a short, 15-minute practice on your own yoga mat each morning that is completely guided by your own bodily intuition. Perhaps it is going to a sweaty hot yoga class or a prenatal class or a yoga-for-real-men class. Any of these choices are totally okay! I think some glimpse of this new awareness is inside of each of us, lying dormant as a super brain power.

If we consider anything that we feel we are experts at, we can probably look back and see all the hard work and practice that got us there, whether it is in sports, academics, music, cuisine, or any other area. If we had known it would take us that long to develop our expertise, we may not have embarked on the journey. The task may have felt too daunting. Meditation can feel like that sometimes, but my recommendation is to just do it. Just start and take one step at a time. Even small bits of meditation practice can offer benefits, and bigger bits can offer benefits that we cannot even fathom. I offer a word of caution, though: meditation practice itself isn't always peaceful and blissful. It can bring up disdain, hardships, and emotional dis-ease. A teacher can be helpful with these and other issues, and in continued practice and study.

Chapter 10

Intuition

Intuition is my all-time favourite super power, so I have left it to the end. It fits nicely after a discussion about mindfulness meditation and yoga because one of the super powers that emerges as a result of those practices — at least for me — was a keener sense of intuition. I think it works because we attune ourselves so deeply through mindfulness and we learn to see things as they are. That attunement, that awareness, that seeing things as they are, *that* is intuition.

I consider myself a relatively intuitive person. I have made life-changing decisions that felt intuitively driven. For example, when I was finishing up my master's degree and contemplating where to go for my PhD, I made a very quick decision to go to Dalhousie instead of the University of British Columbia (UBC). I had been in talks with a potential supervisor at UBC and was arranging a visit to Vancouver to visit my potential lab when a friend of mine at Dalhousie suggested I talk to a new professor there. I resisted, claiming that it was not possible because my boyfriend at the time was unwilling to move to Halifax, but he was willing to move to Vancouver. We had been together for 10 years by that point, so it was a major part of my decision. Eventually I agreed to email the prof at Dalhousie just to appease my friend, but the email exchange quickly turned into a phone conversation with an offer to do my PhD under her supervision. As soon as I hung up, I knew I was going to Dalhousie. It just felt right, despite some major disruptions that I would impose on my personal life.

I couldn't fully explain why it felt right, it just did. I knew that's where I wanted to go. In hindsight, I can imagine several elements at play during this intuitive decision-making process including my sensing my to-be supervisor's eager scientific energy. The science seemed really interesting too, but there was something else I couldn't easily put my finger on. I think it had something to do with her frankness, her experience in competitive sports that I connected and felt familiar with, and maybe the way her mind seemed to work while we were speaking. In any case, I knew without a doubt after we talked that I was going to Dalhousie to work with her, and I never once regretted that decision, even several months later when my boyfriend and I broke up. I knew in my heart that Dalhousie was where I was meant to be.

Intuition also caused me to leave Halifax and move to Toronto. After several months of agony trying to decide if I should go to Toronto with my now-husband, Mike, I woke up one night from a dream where I was living in Toronto. The sentiments in the dream were so powerful, comfortable, and real that I knew that I had to move to Toronto. I had been resisting it out of fear of leaving my comfortable life in Halifax, where I felt like a big fish in a small pond. Thinking about the big city of Toronto terrified me. But something beyond logic, reason, and feeling took over and prompted me to move. It was a sudden, intuitive decision I have never looked back from either, even upon arriving in Toronto with no job and being forced to build my own company all alone. It was one of the hardest times in my life, but it was not a regret-filled time.

I have made many decisions intuitively. I quit my post-doctoral research to join a naturopathic health centre as the life coach. I did my yoga teacher training on a whim. I created a life coach training course, started a company, and broke off several relationships in my life based on these sudden, intuitive, "aha" moments. These experiences are more than just regular decisions. They seem to defy logic, but also seem to incorporate a deeper logic that only becomes obvious in the future. They are full-bodied experiences that speak strongly to me.

Alternatively, I have also felt the consequences of failing to listen to my intuition. When I was applying for my master's degree, I was accepted to both Waterloo and Brock University. On my own, I had decided against Brock and I sent in my letter declining, but after I told my honours degree supervisor about it, she convinced me that I had not made the best choice. I reluctantly retracted my initial letter of decline to Brock, accepted their offer, and went there a few months later to do my master's. In the end, that decision never felt good. In fact, I spent most of my time in St. Catharines in a deep depression. Although I made the best of it and had some great experiences, I can't help but wonder what would have happened if I had trusted my gut and gone to Waterloo.

And I'm not the only one who follows intuition. On one occasion when I was doing my PhD, I went to lunch with a very prominent neuroscientist, Thomas Jessell, co-author of *Principles of Neural Science*, the pre-eminent neuroscience textbook. Someone asked him how he chose his graduate students, appreciating that he would have many excellent people to choose from. His answer was rather simple. He claimed, of course that they needed to demonstrate research capacity and the ability to think intellectually and critically, but at the end of the day, his decision was rather intuitive. I never forgot that. Here was a scientifically minded researcher claiming to make important decisions using his intuition. Clearly, he values it.

Upon further investigation, I have come to realize that many people use intuition to guide decisions. Even big CEOs use it. For example, Apple CEO Tim Cook was quoted in 2011 saying, "I've discovered it's in facing life's most important decisions that intuition seems the most indispensable to getting it right. Intuition is something that occurs in the moment, and if you are open to it, if you listen to it, it has the potential to direct or redirect you in a way that is best for you."[81] Professor Modesto Maidique, Executive Director of the Center for Leadership at Florida International University in Miami, whose research focuses on decision-making, leadership, and executive development, had something similar to say in an article for the *Harvard Business Review*. "The key to effective, intuitive decisions is best conveyed in two wise sayings: 'Know your business' and 'Know yourself.' The sweet spot for business decisions is when both domain knowledge and self-knowledge are high; when you have the

---

81. Tim Cook in Kim Eaton, "Tim Cook, Apple CEO, Auburn University Commencement Speech 2010," *Fast Company*, August 26, 2011, accessed March 30, 2016, http://www.fastcompany.com/1776338/tim-cook-apple-ceo-auburn-university-commencement-speech-2010.

knowledge to shrewdly interpret the facts and the wisdom to steer clear of the biases and destructive emotions that can hinder success."[82] Maidique reinforces the idea that combining emotion and logic results in good intuitive decisions.

Intuition as a Mind-Body Experience

Emotion and logic do combine to foster good decisions, according to psychologist and neuroscientist, Antonio Damasio, who speaks of the combined need for rationality and emotions in decision-making.[83] He describes a perfect example using a patient with frontal-lobe damage. The patient is asked about going out to a restaurant. It starts with him not being able to decide which restaurant to go to because he lists several pros and cons for each of the various options; for example, the emptiness of the restaurant and what that might mean (is the restaurant not good? or is it simply easier to get a table?). The patient ends up in a mental rationality loop that can go on and on for a significant length of time, never reaching a clear decision. Damasio suggests the patient's inability to make a decision is because he fails to get the "lift" from any emotional reaction, and that it is the emotion that allows us to "mark things as good, bad, or indifferent, literally in the flesh."[84]

---

82. Modesto A. Maidique, "Intuition Isn't Just about Trusting your Gut," *Harvard Business Review,* April 13, 2011, https://hbr.org/2011/04/intuition-good-bad-orindiffer/.

83. Antonio R. Damasio, *Descartes' Error: Emotion, Reason, and the Human Brain,* (New York Putnam Publishing, 1995).

84. You can watch Antonio Damasio speak with David Brooks about this phenomenon in this video: https://www.youtube.com/watch?v=IifXMd26gWE.

Damasio has studied several cases of people who suffered from damage largely confined to the prefrontal cortex. In these cases, people suffer from an inability to decide well; for example, they cannot make good decisions with finances and moral and ethical decision-making skills are impaired. These impairments come with no intellectual deficits, no motor deficits, and no other obvious impairment. But it is clear that these individuals have disturbances of emotion that affect their abilities to make decisions, the kinds of decisions I think of as intuitive.

Damasio proposed a theory, called the *somatic marker hypothesis*, to account for how we make decisions based on emotions. He argues that decision-making is a process that is influenced by marker signals that arise in our body (*soma*, you will recall, is the Greek word for "body"). These markers associate positive and negative bodily responses with actions and experiences. Those somatic markers become stored as memories of our past experiences stamped with the bodily sensations associated with emotions. When similar experiences come up in the future, these somatic markers help the brain make decisions relatively quickly, largely because the sensations of emotions (or the markers) arise quickly compared to our appraisal of the situation, which would be largely rationally based. Therefore, these markers allow an individual to use past experiences and past knowledge on an unconscious level. The markers participate in the decision-making process along with the rational input. Many brain areas participate in this process including the orbital frontal cortex (a region within the prefrontal cortex), the amygdala (largely known as the emotion-detecting brain structure), and the somatosensory/insular cortices (areas known for self-awareness, bodily sensation awareness, and possibly disgust and empathy too).[85] As a result of this network of decision-makers, we come up with both the logic and the emotional "lift" Damasio refers to. This lift allows us to make decisions that feel right intuitively.

We don't have to have obvious brain damage to fail to use our emotions in making decisions. Many people struggle with decision-making all the time because they simply don't allow themselves to accept that emotions are part of the process. I too have felt that. Sometimes my body is screaming at me with a set of somatic markers, telling me I don't want to go somewhere because I'm tired and need to rest. Logically, I think I "should" go, and sometimes I do. Other times I listen to the wisdom of those somatic markers and let my body rest.

Now, Damasio may not call this process intuition, but in many ways, this is at least part of what intuition is. Intuition is a high-level neurocognitive information-processing mechanism that exists as a decision-making tool for all humans. Intuition is neither logical nor emotional, but both, as Damasio describes. Intuition is not necessarily a mystical experience that only some people have the luxury of experiencing. We all have access to it by virtue of having a brain, but as with creativity, we don't all recognize or value it as the brain super power that it is.

Psychologist Daniel Kahneman and several other researchers, like psychologist Philip Zimbardo, explain

---

85. Antoine Bechara, Hanna Damasio, and Antonio R. Damasio, "Emotion, Decision-Making and the Orbitofrontal Cortex," *Cerebral Cortex* 10, no. 3 (2000): 295–307.

where intuitive decision-making can be more successful. One such situation exists when the decision maker has a great deal of expertise in the area. This is why business people can make great decisions based on gut, or intuitive feelings, but only when they are substantially experienced. Stories of this kind exist among many kinds of experts: car mechanics, nurses, firefighters, and ER doctors, too. One story that I love, told by Kahneman himself, was of a firefighter who was on the roof of a building with his team. All of a sudden he yelled for everyone to run and get out of the building. Within seconds of exiting, the building (including the roof they had been on) collapsed. When probed as to why he had the sudden instinct to run, the firefighter couldn't explain it. He said it was just a gut feeling that he had acted on. Eventually, this intuition was chalked up to his being able to detect heat beneath his feet, so he knew instinctively spread and the structure of the building would be compromised. He made a decision based on that led to a gut feeling reaction. This is so fast that it could not possibly be conscious level and still be useful. If the forced to stand there, logically thinking about the sensations in his feet, it would likely be too late for him and for his team. By the time his consciousness kicked in, the building might have already collapsed.

Neuroscience of Intuition

There is no solid theory on the neurobiology of intuition, largely because intuition itself is difficult to define. Defining intuition is like attempting to define love or happiness, both of which also lack a solid neurobiological theory. Despite this, several areas of the brain have been implicated in the experience of intuition and have been addressed in different research paradigms. Some researchers, like Kahneman, have attempted to define and study intuition using fast- and slow-thinking models.[86] Kahneman refers to fast thinking as more intuitive and automatic than slow thinking, relying on available information to make quick decisions, and largely unconsciously.

Others have tried to define and study intuition based on implicit memory, or memories that we can't explicitly claim to have, but that are based on acquired knowledge with time, practice, and routine, things like riding a bike, driving a car, or brushing our teeth. We rely on skill-based memories like these in our daily life, but we can't always say when or how we learned them. In some ways, these types of memories feel analogous to "intuitive knowing." The same is true when I consider skills in various sports, like ultimate or skating. I can't describe how or when I learned to do things, but I must have formed some memory of the skill in order to replicate it during the games. I don't think about throwing a disc or putting one skate in front of the other, but these things happen as if my body just knows what to do.

Despite the lack of a solid definition, it still seems clear that there is some kind of information-processing or decision-making process occurring within the brain that allows us to experience, know, learn, and remember things beyond our awareness. These processes seem to

---

86. Daniel Kahneman, *Thinking, Fast and Slow* (New York: Farrar, Straus and Giroux, 2011).

drive our behaviours and are rooted in our brains. The brain areas implicated in this process are numerous, likely because this high-level thinking requires many information-processing and decision-making brain structures to merge information and allow "intuition" to express itself. At the top of the brain pyramid are several cortical structures like the orbitofrontal cortex, a region within the frontal cortex. In a 2014 scientific study,[87] researchers Ninja Horr, Christoph Braun, and Kirsten Volz propose that the orbitofrontal cortex integrates the "gist" of the information when information is incomplete. For example, we might see a partial image of someone or something and be able to get the gist of who or what is in the image by using only partial information.

The feeling associated with the guess we make seems to be related to the activation of our orbitofrontal cortex based on the information it does have available. The precuneus, another cortical region in the frontal lobes, also appears to have a role. This is apparent in expert game players who, when quickly or intuitively generating a best-next move, have a corresponding activation in the precuneus.[88] The researchers in this study add that the precuneus seems to send its information to the dorsolateral prefrontal cortex and then to the caudate so that action can arise based on that guess.

The caudate, a subcortical structure involved in memory of motor skills, including implicit memories,

---

87. Ninja K. Horr, Christoph Braun, and Kirsten G. Volz, "Feeling Before Knowing Why: The Role of the Orbitofrontal Cortex in Intuitive Judgments— An MEG Study," *Cognitive Affective and Behavioral Neuroscience* 14, no.4 (2014): 1271–1285.

88. Xiaohonog Wan et al., "The Neural Basis of Intuitive Best Next-Move Generation in Board Game Experts," *Science* 331, no. 6015 (2011): 341–6.

also seems to be involved.[89] Imagine you were driving down a road you had travelled a few times but that you were not entirely familiar with. You might think you need to turn right to get to your destination, but then your GPS alerts you to turn left. You aren't entirely sure whether to trust the GPS because something inside you is telling you to do the opposite. That pull to turn right is probably mediated by activity within the caudate because it previously coded your movement as a memory. The caudate likely has to do with providing the stamp of familiarity to any body sensations that arise. If turning right were correct and you followed your "intuition," we might very well say you followed your caudate. The activity of the caudate might be part of the intuitive decision-making otherwise described as our "body's wisdom" or "inner wisdom". Indeed, the caudate is inside us, turning what some people think of as "flaky" into what others call biology.

As already mentioned, we also know that intuition appears to be best used by people with significant amount of experience within a domain. Master chess players, emergency-room doctors, athletes, and seasoned CEOs all seem more apt to use intuition to guide decisions than novices. This expertise might recruit the help of another well-known structure, the hippocampus. The hippocampus is, as discussed earlier, involved in memory, but it is particularly involved in memory of space and where we are in space, as in awareness of our surroundings and our environments. The more familiar

---

89. Xiaohonog Wan et al., "Developing Intuition: Neural Correlates of Cognitive-Skill Learning in Caudate Nucleus," *The Journal of Neuroscience* 32, no. 48 (2012): 17492–17501.

we are with our surroundings, the better our hippocampus has coded it as "place cells." The familiarity stamping done by the hippocampus could help us sense when something is out of place, even if we cannot explicitly state what it is. A strong and healthy hippocampus might be the key component in a master chess player's familiarity with the board, an emergency doctor's ability to contextualize ailments as symptom profiles quickly, an athlete's hyper-familiarity with their field of play, or a CEO's keen perception of the market or business landscape.

The hippocampus is also part of the subcortical system known as the limbic system, which was discussed earlier as being the emotional circuitry of the brain. Another key player in that system is the amygdala, which serves to quickly and subconsciously process information and give rise to an arousal experience, often in the form of an alert. The amygdala has also been implicated in the neural basis of intuition. The amygdala may represent at least part of the emotional "lift" described in reference to the somatic marker hypothesis. It then communicates its interpretation, along with other subcortical information, back up to the cortical regions within the frontal cortex.

Collectively, these (and likely other) regions all seem to play a role in what we experience as intuition. There is also evidence that evaluating risk might require a neurobiological "gut feeling." When a group of researchers examined neural activation when participants estimated hazards to be risky, the researchers found activation in the subcortical structures of the medial thalamus and caudate nucleus, but also in cortical regions of the anterior insula, cingulate cortex, and further prefrontal and temporo-occipital areas.[90]

There are many ways we could tackle trying to understand the neurobiology of intuition, and indeed that research is likely to continue. Intuition is a brain-based decision-making process incorporating thoughts, body sensations, and several key areas within the brain that are directing all of this. Putting this together allows us to make use of the brain super power.

How to Use Intuition

There are countless books written on how to develop, find, use, or improve your intuition. I'm not entirely sure how helpful those books are. But what seems clear is that intuition is an experience-based skill that requires practice and expertise. In addition to practice, refining the skill of intuition requires feedback on the accuracy of the intuition. There are some intuitive decisions that we may never get to test. For example, many years ago I was on a canoe trip with some friends. There were designated camping spots along the water where we were canoeing. We arrived at a site late in the day, and we knew the next one was a couple of hours away. As we set foot on the land, we all felt an incredible sense of discomfort. It was eerie. None of us spoke as we began to unload, but it eventually became clear that we were all feeling the same discomfort at this site. Unanimously, we agreed to leave and head to the next location. We never did find out if there was anything to our gut feelings and in some ways

---

90. Uwe Herwig et al., "Neural Correlates of Evaluating Hazards of High Risk," *Brain Research* 1400 (2011): 78–86.

we were happy to not know. Not 10 minutes later in our canoe, we happened upon a big moose staring from the bank of the very narrow river we were paddling. Would that moose have trampled us in that other campsite? Who knows. The situation did not allow us to ever know whether our intuition was right or wrong. All we got to learn was that following our intuition did not harm us, and we had a wonderful stay in the next campsite. That's something to be thankful for.

Finding out if our gut feeling or intuition is correct does require us to receive feedback. Although many situations are like my canoe trip, where we'll never know for sure if our intuition steered us correctly, there are many situations where we can assess the accuracy. Indeed, some of us will be guided to assess and affirm our own intuition, and the practice will serve us well in the ability to use our intuition wisely. We can begin our assessment in simple ways, like listening to our inner compasses when we feel like going in one direction but a GPS tells us to go the other, and then assessing our accuracy. When we have a gut feeling on an investment decision, we can invest and then assess the result. When we get an intuitive sense that something is wrong with a friend, we can call them up, ask, and assess the accuracy. Those are just some of the ways we can get positive confirmation. For the situations where we get no confirmation, perhaps we can be thankful if no harm comes to us when we follow our instincts and maybe just leave it at that. If we can learn to trust our intuition after practice, there will be good reason to trust more as we gain a deeper awareness of how our bodies, minds, and brains react to different situations.

Feedback is important when it comes to relying on our intuition, according to psychology research. We can also get feedback during the activities we engage in that make us experts, which often involve our jobs or specific tasks within our jobs. We also develop expertise if we become well-trained athletes within a specific sport. We may become relationship experts, guidance experts, gaming experts, dog-behaviour-analysis experts, writing experts, music experts... All kinds of expertise can be developed in our day-to-day living. By investing in activities that we enjoy, we ultimately end up being able to make quick decisions and process information more quickly at an unconscious level than a novice would be able to. Is that intuition? At the very least, it's one expression of it and, as such, a very important and practical brain super power.

Chapter Assignment: Test Your Intuition

The mindfulness practice from before will help you tune into the body sensations associated with different experiences. Some of those body sensations are signals that you can use to make decisions, just as described by Antonio Damasio when referring to somatic markers. Start with an awareness exercise: Allow yourself to check in regularly with what your body sensation signals might be. When you're about to make a decision, ask yourself what you feel in your body. Listen to see if anything comes up. Take note of these sensations in a journal.

If you keep track of the times when body sensations are associated with decisions over several weeks (or months or even days, depending on how attuned you are), you may start to observe a pattern. Certain signals might be associated with when you should say "no" and others may be associated with when you should say "yes." You may also start to notice somatic markers associated with failing to listen to either of those.

Eventually, you might want to really experiment with your intuition in situations that don't carry a lot of consequences.

Try actively going against your intuition and then observing what happens!

Chapter 11

Creativity

Our brain is creative, yet we don't all recognize this as the super power that it is. When I lecture on creativity, I often start by asking who in the room thinks of themselves as creative. Though it depends on the audience, most people don't raise their hands. Even when some hands do raise, many still sit there unable to believe they are creative. But while few people consider themselves creative, creativity is in our nature, in our genes, and in our brains! Look around ... all the "stuff" around us did not emerge without the incredible human brain's capacity to create. Buildings did not erect themselves. Books did not write themselves. Courses did not design themselves. Canvasses did not paint themselves. Fashion did not walk itself down the runway. All of these are examples of the human brain expressing its creativity. We can even see the creative capacity in everyday life: relationships and friendships, jobs and job opportunities, vacations, ideologies, and strategies are all created regularly. Babies and houses and lifestyles are all created. New sports, teams, businesses, non-profits, all kinds of events, dinners and dinner parties, weddings, celebrations of life, latte art, all electronics, new ways of thinking... all created through the power of our brains.

Creativity is natural in humans, even on the basic level of sexual reproduction — where we literally create human beings — but creativity is much deeper than that. In fact, before being born, and with the help of our mothers, we exert a significant amount of energy in *creating* connections among neurons. These connections create a complex brain system, with some of the most interesting connections happening in our frontal cortices. This part of the brain is quite distinct in humans, relative to other animals, and is often considered responsible for many of the incredible abilities that we tend to consider "human," including our creativity.

Neuroscience of Creativity

In a great TED talk, medical doctor and researcher Charles Limb[91] describes his curiosity about creativity. He began to study musicians' abilities to improvise in both jazz and rap by collecting images with fMRI. He found that when musicians engaged in improvisation, brain areas otherwise known to be involved in self-monitoring (dorsolateral prefrontal cortex and lateral orbital cortex) were deactivated while, at the same time, an area involved in self-expression (medial prefrontal cortex) was ramped up. Limb explains these neural findings as representative of a musician's ability to shut down inhibitions associated with self-monitoring and let the inner voice shine through. In other words, Limb found

---

91. Charles J. Limb, "Your Brain on Improv," *TED Talk* (November 2010), http://www.ted.com/talks/charles_limb_your_brain_on_improv?language=en

neural evidence for an inner critic of sorts, which may be what silences our inner creative. It's interesting that the medial prefrontal cortex (particularly the area on the right side of the brain) pops up, or rather lights up, in this study, largely because it is also known to be involved in the production of original ideas.[92]

Similarly, researchers Shamay-Tsoory and colleagues argue that creative cognition may be interfered with by the dominant language areas of the brain.[93] If these areas becomes damaged, our creativity is unleashed. Actually, there are several neurological conditions in which creativity is facilitated through damage. This paradox happens in some forms of epilepsy and Alzheimer's disease, and patients can develop heightened creativity. It's kind of like killing the inner critic. The inner critic might also involve another area of the brain, the precuneus, which is typically activated during self-consciousness and some types of memory (episodic memory). This area isn't in the frontal lobes, but rather in the cortex at the top of the brain, known as the parietal lobes. Of further interest, researchers Takeuchi and colleagues[94] found the precuneus to have reduced activity in creative individuals, further suggesting a quiet "inner critic" associated with greater creativity.

---

92. Chalres J. Limb and Allen R. Braun, "Neural Substrates of Spontaneous Musical Performance: An fMRI Study of Jazz Improvisation," *PLoS ONE* 3, no. 2 (2008).

93. Simone G. Shamay-Tsoory et al., "The Origins of Originality: The Neural Bases of Creative Thinking and Originality," *Neuropsychologia* 49, no. 2 (2011): 178–185.

94. Hikaru Takeuchi et al., "Cerebral Blood Flow During Rest Associates with General Intelligence and Creativity," *PLoS* 6, no. 9 (2011): e25532. doi: 10.1371/journal.pone.0025532

Unleashing the Creative Brain

How can we apply this information to our own brains? Well, let's assume that those of us who do not normally put up our hands when asked "who here is creative?" are now convinced that maybe, *just maybe*, we all have a creative potential, even if it's suppressed by an inner critic. Learning to quiet the inner critic and letting our inner creative speak up might serve us well. We're going to attempt this with a reframing exercise similar to those we did in earlier chapters about negative thoughts and failures.

Awareness. List the creative activities you engage in during any given day or week. Similarly, a gratitude journal would work to highlight creative activities too.

Reframing. Anyone who would not raise his or her hand in a room when asked "who here is creative?" should consider reframing that thought process. When you hear yourself (or your brain) say anything remotely similar to "I'm not creative," this is your cue to argue with your brain. Explain to it all the reasons that you *are* creative. Rewrite the script in your brain and let your brain change itself. Next time someone asks "who here is creative?", put up your hand and let your body also reframe your mind.

Practice. One practice is the paper clip exercise, which fosters one aspect of creativity: divergent thinking. For this experiment, set a timer for one minute then list all of the ways a paper clip can be used for something other than keeping paper together. Kids can usually come up with about 100 in only a couple of minutes. Adults ... well, we need practice. We're lucky if we can come up with 15 without much practice. Their inner critic quickly stamps out nonsensical ideas and literally truncates the creative process. Adults, therefore, need much more practice to override a tendency we developed (or rewired) that got us out of the habit of being creative. Each day practise this exercise with a different object. Over time, it will get easier and easier as your mind expands and your brain changes.

In many ways, our brain's ability to create is its greatest asset and the limits of our creativity are unknown to many of us. Few of us could have predicted 100 years ago all that our brains have created today. With this powerful creativity we can innovate and create solutions to many of our own personal problems but also create solutions to global problems. But, we have to believe in our own brain's creative potential. It's there. Ready to be unleashed.

Chapter 12

Summary and Conclusion

Our brains are already super powerful and so are we. The brain can trick us, heal us, help us, hinder us, be aware for and of us, create for us, and intuit for us. It is incredibly smart, wise, and insightful. How best can we use this powerful tool that we have? On paper, it's quite simple:

Rest, relax, and rejuvenate the brain. The brain cannot handle most of the stress we throw at it if we expect a great deal from it. Sure, our brains are incredibly resilient and can endure much of what we do, but in order for us to achieve a satisfied and self-actualized existence, we need to give our brains a break. The brain needs to sleep and to relax its frontal lobes and emotional centres at various times during the day. It needs proper nourishment of food and water so it can function properly, develop neurochemicals, and have enough resources to do its magic.

Be mindful. One way of achieving rest and relaxation is by engaging in mindfulness practices like meditation and yoga. But in addition to offering those fundamentally nourishing elements to our self care, mindfulness practices also offer us the ability to develop significant wisdom, inner intelligence, and deeper self-awareness. This kind of knowledge is so fundamental that it has been sought after for thousands and thousands of years, yet in the West we have only recently adopted these "new fads." Only yoga and meditation are not fads. They are a true practice of self-actualization and ultimate knowing.

Think positively. Train ourselves to think positively and optimistically under most circumstances (with caveats in mind). We fare much better when we do this: our health is better, successes happen, and we feel better. Of course, this should never be at the expense of suppressing our emotions. Negative emotions need to be acknowledged and then, when appropriate, transformed into something helpful, actionable, useful, and ultimately more positive. In the process we can give gratitude and positive feedback and be more receptive when others give gratitude and feedback to us.

Embrace failures, mistakes, and wrongness as natural and important contributors to success. We will benefit if we acknowledge and celebrate failures, not hide them. We can recognize that we are human, and it's unrealistic to expect perfection. Delusion does not serve us.

Know what motivates us. Motivate ourselves with intrinsically interesting activities when we can. Reward ourselves for our actions when we are working toward something that is not intrinsically motivating.

Check in. The brain also needs to check in on itself regularly. We achieve this by talking with others and engaging in our own critical assessment of the thoughts our brains produce. Talking out loud to others is often helpful to check if we are making sense, and so is going through cognitive reframing exercises to weigh evidence against our negative thinking. This can help direct the power of the mind in a more productive and meaningful manner. This also allows us to check in on our happiness and optimism levels and alter thinking patterns, should we find our brain going off track. These are practices, just like mindfulness, that need to become regular and routine. SMARTER goals are helpful in establishing and reinforcing this pattern. So is having a strong community of people whose minds we aim to model.

Engage in higher thinking. We need to trust in the brain's power to provide quick, intuitive information. Our brains are creative, and we can and should unleash the creative power. The brain is wired for these experiences and for these elements of our existence. To deny our brains these not only truncates our potential, but deprives us of a meaningful and full life.

Practise using intuition.

The reality of doing all of these well is the challenge, but it's one that I hope we are all more inspired to accept, having learned a bit more about our own brains. I believe that all of this boils down to self-awareness, as both a practice and a goal. The manner by which we get there is entirely unique. Neuroscience, psychology, Wicca, yoga, mindfulness, insight meditation, having a baby, owning a dog, moving from Winnipeg, contemplating anxiety or body image or depression, teaching, learning, dreaming, and loving from one particular manner (mine!), but I'm deeply in awe and support of each and everyone else's methods too.

I have nothing more profound to leave you with. My hope is that you have found epiphany, awareness, knowledge, and insight within your own brain that will serve you from this point onward. The point of this book is to help you find your own inner guide. The content is here to point you in the direction of your true self. In fact, it's time for me to go on a new journey too. In the process of finishing this manuscript and publishing it, I got pregnant and had a beautiful baby boy. The first few months after he was born were among the most challenging times of my life. Nothing bad happened. Everyone is well and healthy. It has been another incredible self-reflecting journey, albeit a trying one. A few months in, I'm only now emerging from a cloud of confusion and culture shock with an altered self-identity. This new identity has taught me so much already, allowing me to revisit many elements of this book. In some ways, my brain, body, and mind feel like they have all been completely rewired to be a mom. This experience serves to remind me that we are never done. We are always evolving and transforming, whether we want to or not, and whether we acknowledge it or not. Some parts of that evolution and transformation will be welcomed. Some not so much. But in the end, we, our minds, bodies, and brains are ever changing, and all we can hope for is to be aware of it happening and to enjoy the ride.

I know there is an incredible power inside each of us that just needs to be realized, actualized, and materialized. That thought drives me, and hopefully it will drive you too. I invite you to continue to explore, discover, learn, and to be aware. If desired, please share those experiences with me or with others around you. It really is more fun to do this with others. Thank you for coming on this journey with me.

Namaste. The light in me honours the light in you.

Acknowledgements

There are many, many people to thank for helping get this book published! Saying "thank you" to everyone gives me such a tremendous sense of inner happiness (just as I describe in the book about the practice of gratitude). Thank you, everyone!!!

Specifically, I want to thank the many people who helped in the publishing process. Thank you to my copy editor, Kate Unrau, for taking an interest in the manuscript and providing me with the confidence to get it done, not to mention putting up with long delays caused by my having a baby in the middle of the project. Also, thank you for helping with the crowdfunding campaign and, of course, for your excellent editing help! Thank you to my proofreader, Holly Warren, for taking care of all that detail-oriented jazz that would cause my brain to explode. Thank you to my publisher, Kathryn Willms, for being so understanding, patient, and positive throughout this process, and thank you for all the things you did behind the scenes that I don't even know about! Thank you to Greg Ioannou for taking an interest from the beginning and deciding to publish this manuscript. Thank you to my crowdfunding platform, Pubslush (which is sadly no longer in operation, but which was a great help to me). Thank you to my designer, Chang Baek, for the great illustrations and being so easy to work with. Thank you to my marketer, Emily Niedoba, for some early branding work. Finally, thank you to my fellow Iguana author,

Paul Dore, who led the way and provided me with helpful knowledge and inspiration.

I also want to thank a few key coaches — friends, colleagues, and staff. Allison McDonald, thank you for your unwavering support of all things related to the production of this book and the delivery of much of this content through the many courses we have run together. Your physical and emotional support has been instrumental in getting this done! Vanessa Vella, thank you for reviewing this manuscript and for providing some coaching for the crowdfunding campaign. Thank you for your other administrative support that gave me time and space to focus on getting this done! Amanda Celis, thank you for helping with some of the early scientific literature searches. Dawn Stevens, thank you for reviewing a version of this book way back in the day — you may not even remember! — and for your helpful comments when this manuscript served as a manual for the life-coaching course. Thank you Louis Lakatos for your support of my eccentric work, including the idea of this book. Thank you Olivia Ragoza for signing up for the life-coaching course through the crowdfunding campaign!

Of course, I also want to thank other important members of my wonderful community, including my partner, Mike Lovas, who has given me space, encouragement, and pushes along the way. I love all of our conversations that continue to spark, fuel, and challenge many of my ideas that come out in this book. Thank you to my amazingly supportive mother, Avis Johnston, for inspiring in me the desire to do things my way, to be strong, and to persevere, all of which were necessary components for this book to happen. Thank you also for helping me (literally) deliver these ideas through my life-coaching course. More recently, thank you for taking care of our little baby Ashar and showering him with so much love and for teaching him so many new things while I got to finish this book. Thank you to my father, Gord Wintink, for always fostering and supporting my interest in psychology and neuroscience and also funding me in many ways! A big huge thanks to my soul sister Lindsey White and her partner Bronwyn Whyte for all your support and for challenging me to think deeper about pretty much all things life related. A special thanks to Lindsey for being energetically connected with me as we both seek to create more openness and sharing in this world, one important aspect of this book. Thank you also for the many facets of support from my second parents, John and Sandy Lovas. Thank you to my Gigi and Grandma Wintink for much support, emotionally and financially. And thank you to my other friends and family for their extra generous support through the crowdfunding campaign: Kathi Wintink, Vivian and Wally Remple, Andrea Darnborough, John and Sandy Lovas, Kristan Wintink, Kristen Guy, Nicole Herschenhous and Dave Lovas, Laurel and Nick Murphy, Allison McDonald, Amanda Celis, and Jennifer Chan. Jenn, you get extra extra thanks for being an inspiring woman entrepreneur and fellow shift disturber. Thank you also to everyone who supported my crowdfunding campaign by pre-ordering a book!

A few others also deserve a special thanks for helping me along to a place where I could even imagine voicing my opinion, never mind publishing a book. Thank you to my "smarty-pants" colleagues and friends: Lisa Fiorentina,

Marsha Penner, Sandra Wiebe, Sarah Johnson, and Sara Burke. Thank you to my pre-Google, information-request buddy, Alicia (Davis) Harder and the other "Kalynchicks". Thank you to all those who read, comment, or react to my Facebook posts. Thank you to my PhD supervisor, Lisa Kalynchuk, for giving me so much freedom and confidence in the lab and for providing a foundation for me to explore a new way of teaching neuroscience.

Last, but certainly not least, is a huge thank you that I owe to my amazing son Ashar. First of all, thanks for waiting a few extra days to show up so I could finish a draft. But thank you also for challenging me beyond belief in a way that is turning out to be one of the hardest and most rewarding experiences of my life, a lot of which I am writing and talking about already, which will likely show up in another book.